THE CASE FOR INTERNATIONAL SHARING OF SCIENTIFIC DATA: A FOCUS ON DEVELOPING COUNTRIES

PROCEEDINGS OF A SYMPOSIUM

Kathie Bailey Mathae and Paul F. Uhlir, *Editors*

Committee on the Case of International Sharing of Scientific Data: A Focus on Developing Countries
Board on International Scientific Organizations
Board on Research Data and Information
Policy and Global Affairs

In collaboration with the Committee on
Freedom and Responsibility in the Conduct of Science
International Council for Science

NATIONAL RESEARCH COUNCIL
OF THE NATIONAL ACADEMIES

THE NATIONAL ACADEMIES PRESS
Washington, D.C.
www.nap.edu

THE NATIONAL ACADEMIES PRESS   500 Fifth Street, NW   Washington, DC 20001

NOTICE: The project that is the subject of this report was approved by the Governing Board of the National Research Council, whose members are drawn from the councils of the National Academy of Sciences, the National Academy of Engineering, and the Institute of Medicine. The members of the committee responsible for the report were chosen for their special competences and with regard for appropriate balance.

This study was supported by the National Science Foundation (Award No. OISE-0614728 and OGI-1040898). Any opinions, findings, conclusions, or recommendations expressed in this publication are those of the authors and do not necessarily reflect the views of the organizations or agencies that provided support for the project.

International Standard Book Number 13: 978-0-309-30157-2 (Book)
International Standard Book Number 10: 0-309-30157-2 (Book)

Additional copies of this report are available from the National Academies Press, 500 Fifth Street, NW, Room 360, Washington, DC 20001; (800) 624-6242 or (202) 334-3313; http://www.nap.edu.

Copyright 2012 by the National Academy of Sciences. All rights reserved.

Printed in the United States of America

THE NATIONAL ACADEMIES
Advisers to the Nation on Science, Engineering, and Medicine

The **National Academy of Sciences** is a private, nonprofit, self-perpetuating society of distinguished scholars engaged in scientific and engineering research, dedicated to the furtherance of science and technology and to their use for the general welfare. Upon the authority of the charter granted to it by the Congress in 1863, the Academy has a mandate that requires it to advise the federal government on scientific and technical matters. Dr. Ralph J. Cicerone is president of the National Academy of Sciences.

The **National Academy of Engineering** was established in 1964, under the charter of the National Academy of Sciences, as a parallel organization of outstanding engineers. It is autonomous in its administration and in the selection of its members, sharing with the National Academy of Sciences the responsibility for advising the federal government. The National Academy of Engineering also sponsors engineering programs aimed at meeting national needs, encourages education and research, and recognizes the superior achievements of engineers. Dr. Charles M. Vest is president of the National Academy of Engineering.

The **Institute of Medicine** was established in 1970 by the National Academy of Sciences to secure the services of eminent members of appropriate professions in the examination of policy matters pertaining to the health of the public. The Institute acts under the responsibility given to the National Academy of Sciences by its congressional charter to be an adviser to the federal government and, upon its own initiative, to identify issues of medical care, research, and education. Dr. Harvey V. Fineberg is president of the Institute of Medicine.

The **National Research Council** was organized by the National Academy of Sciences in 1916 to associate the broad community of science and technology with the Academy's purposes of furthering knowledge and advising the federal government. Functioning in accordance with general policies determined by the Academy, the Council has become the principal operating agency of both the National Academy of Sciences and the National Academy of Engineering in providing services to the government, the public, and the scientific and engineering communities. The Council is administered jointly by both Academies and the Institute of Medicine. Dr. Ralph J. Cicerone and Dr. Charles M. Vest are chair and vice chair, respectively, of the National Research Council.

www.national-academies.org

COMMITTEE ON THE CASE FOR INTERNATIONAL SHARING OF SCIENTIFIC DATA: A FOCUS ON DEVELOPING COUNTRIES

FAROUK EL-BAZ (*Chair*), Boston University
BARBARA ANDREWS, University of Chile
ROBERTA BALSTAD, Center for International Earth Sciences
JOHN RUMBLE, JR., Information International Associates, Inc.
WILLIAM WULF, University of Virginia
TILAHUN YILMA, University of California, Davis

Staff

KATHIE BAILEY MATHAE, Study Director
PAUL F. UHLIR, Study Director
LYNELLE VIDALE, Program Associate
CHERYL WILLIAMS LEVEY, Senior Program Assistant

BOARD ON INTERNATIONAL SCIENTIFIC ORGANIZATIONS

Cutberto Garza, MD (IOM), *Chair*, Boston College
Marvin Geller, Stony Brook University
Daniel Goroff, Alfred Sloan Foundation
Priscilla Grew, University of Nebraska State Museum
Melinda Kimble, United Nations Foundation
Dennis Ojima, Colorado State University
Kennedy Reed, Lawrence Livermore National Laboratory
John Rumble, Jr., Information International Associates, Inc.
Karen Strier (NAS), University of Wisconsin
Tilahun Yilma (NAS), University of California, Davis

EX OFFICIO
Roberta Balstad, *Retired*
Michael Clegg (NAS), University of California, Irvine
Dov Jaron, Drexel University
J. Bruce Overmier, University of Minnesota

BOARD ON RESEARCH DATA AND INFORMATION

Francine Berman, Cochair, Rensselaer Polytechnic Institute
Clifford Lynch, Cochair, Coalition for Networked Information
Laura Bartolo, Kent State University
Philip Bourne, University of California, San Diego
Henry Brady, University of California, Berkeley
Mark Brender, GeoEye Foundation
Bonnie Carroll, Information International Associates
Michael Carroll, Washington College of Law, American University
Sayeed Choudhury, Johns Hopkins University
Keith Clarke, University of California, Santa Barbara
Paul David, Stanford Institute for Economic Policy Research
Kelvin Droegemeier, University of Oklahoma
Clifford Duke, Ecological Society of America
Barbara Entwisle, University of North Carolina
Stephen Friend, Sage Bionetworks
Margaret Hedstrom, University of Michigan
Alexa McCray, Harvard Medical School
Alan Title, Lockheed Martin Advanced Technology Center
Ann Wolpert, Massachusetts Institute of Technology

EX OFFICIO
Robert Chen, Columbia University
Michael Clegg, University of California, Irvine
Sara Graves, University of Alabama in Huntsville
John Faundeen, Earth Resources Observation and Science Center
Eric Kihn, National Geophysical Data Center (NOAA)
Chris Lenhardt, Oak Ridge National Laboratory
Kathleen Robinette, Air Force Research Laboratory
Alex de Sherbinin, Columbia University

PREFACE AND ACKNOWLEDGMENTS

Scientific research and problem solving are increasingly dependent for successful outcomes on access to diverse sources of data generated by the public and academic research community. Global issues, such as disaster mitigation and response, international environmental management, epidemiology of infectious diseases, and various types of sustainable development concerns, require access to reliable data from many, if not all, countries. Digital networks now provide a near-universal infrastructure for sharing much of this factual information on a timely, comprehensive, and low-cost basis. There also are many compelling examples of data sharing in different research and application areas that have yielded great benefits to the world community, although many more could be similarly facilitated.

Many countries that are members of the Organisation for Economic Co-operation and Development (OECD) and some emerging economies already have implemented national policies and programs for public data management and access, while others are in the process of developing them. Nevertheless, many developing countries do not have formal mechanisms in place. The topic of "data sharing" is broad and complex, and developing countries have different infrastructure, human resource, and access needs that must be addressed. (For purposes of this report, "developing" countries are defined as non-OECD countries, recognizing that there is a broad range of economic development among the non-OECD nations.)

There are various specific barriers to the access and sharing of scientific data collected by governments or by researchers using public funding. Such obstacles include scientific and technical, institutional and management, economic and financial, legal and policy, and normative and sociocultural barriers, as well as limitations in digital infrastructure. Some of these barriers are possible to diminish or remove, whereas others seek to balance competing values that impose legitimate limitations on openness. Despite such challenges, however, there could be much greater value and benefits to research and society, particularly for economic and social development, from the broader use and sharing of existing factual data sources.

Many researchers in developing countries, in particular, lack the norms and traditions of more open data sharing for collaborative research and for the development of common research resources for the benefit of the entire research community. Moreover, the governments in many developing countries treat publicly generated or publicly funded research data either as secret or commercial commodities. Even if governments do not actively protect such data, many lack policies that provide guidance or identify responsibilities for the researchers they fund concerning the conditions under which researchers should make their data available for others to use. Finally, developing countries frequently do not have data centers or digital repositories in place to which researchers can submit their data for use by others. In those cases where such repositories do exist, they tend to be managed as black archives—that is, not open to most researchers or the general public.

Because of the importance of data access and sharing in the developing world, an ad hoc committee of the Board on International Scientific Organizations (BISO) and the Board on Research Data and Information (BRDI), in consultation with the Committee on Freedom and Responsibility in the Conduct of Science (CFRS) of the International Council for Science (ICSU), organized a 2-day international symposium in Washington, D.C., on April 18–19, 2011. The main objective of the symposium was to gain better understanding of the data access and sharing situation in the developing world, with a focus on barriers, opportunities, and future actions.

Part One of the proceedings addresses the following questions: Why is the international sharing of publicly funded scientific data important, especially for development? What are some examples of past successes, and what are the types of global research and applications problems that can be addressed with more complete access to government data collections and government-funded data sources?

Part Two provides an overview of the status of public data access internationally, particularly in developing countries. Part Three explores the principal barriers and limits to sharing public data across borders. Finally, Part Four discusses the rights and responsibilities of scientists and research organizations in providing and getting access to publicly funded scientific data. It also provides some insights on how international scientific organizations, government agencies, and scientists can more successfully improve sharing of publicly funded data to address global challenges, particularly in less economically developed countries.

This proceedings contains edited versions of the symposium presentations. As such, they vary in length, formality, and style. Some are more scholarly than others. In addition, language usage varies, since many of the international presenters are nonnative English speakers.

The proceedings is intended primarily for government policy makers, researchers in the developing world, and managers in public and private institutions that fund research and development activities in developing countries. We hope it will enrich their understanding of the importance of data access and reuse from publicly funded research, especially in the developing world, and that it will advance discussions about future actions.

This volume has been reviewed in draft form by individuals chosen for their technical expertise, in accordance with procedures approved by the National Research Council's Report Review Committee. The purpose of this independent review is to provide candid and critical comments that will assist the institution in making its published report as sound as possible and to ensure that the report meets institutional standards for quality. The review comments and draft manuscript remain confidential to protect the integrity of the process.

We wish to thank the following individuals for their review of *The Case for International Sharing of Scientific Data: A Focus on Developing Countries Proceedings of a Symposium:*

> William Anderson, Praxis 101; Peter Arzberger, University of California, San Diego; R. Stephen Berry, University of Chicago; Anita Eisenstadt, National Oceanic and Atmospheric Association; and Kamran Naim, University of Tennessee.

Although the reviewers listed above have provided constructive comments and suggestions, they were not asked to endorse the content of the individual papers. Responsibility for the final content of the papers rests with the individual authors.

> *Farouk El-Baz*, Chair, Committee on the Case for
> International Sharing of Scientific Data: A Focus on
> Developing Countries
>
> *Kathie Bailey Mathae*, Director, Board on International
> Scientific Organizations
>
> *Paul F. Uhlir*, Director, Board on Research Data and
> Information

CONTENTS

1. Welcoming Remarks   1
    Charles Vest, National Academy of Engineering
    United States

PART ONE: SETTING THE STAGE   3

2. Background and Purpose of the Symposium: Historical Perspective   4
    Farouk El-Baz, Boston University
    United States

3. Why Is International Scientific Data Sharing Important?   7
    Dr. Atta-ur-Rahman, UNESCO Science Laureate
    Pakistan

4. Discussion of Part One by Symposium Participants   15

PART TWO: STATUS OF ACCESS TO SCIENTIFIC DATA   17

5. Overview of Scientific Data Policies   18
    Roberta Balstad, Columbia University
    United States

6. Implementing a Research Data Access Policy in South Africa   21
    Michael Kahn, University of Stellenbosch
    South Africa

7. Access to Research Data and Scientific Information Generated with Public Funding in Chile   24
    Patricia Muñoz Palma, National Commission for Scientific and Technological Research
    Chile

8. The Management of Health and Biomedical Data in Tanzania: The Need for a National Scientific Data Policy   27
    Leonard E. G. Mboera, National Institute for Medical Research.
    Tanzania

9. The Data-Sharing Policy of the World Meteorological Organization: The Case for International Sharing of Scientific Data   29
    Jack Hayes, U.S. Permanent Representative to the World Meteorological Organization
    United States

10. Discussion of Part Two by Symposium Participants ... 32

PART THREE: COMPELLING BENEFITS ... 37

11. Developing the Rice Genome in China ... 38
    Huanming Yang, BGI
    China

12. Data Sharing in Astronomy ... 41
    Željko Ivezić, University of Washington
    United States

13. Sharing Engineering Data for Failure Analysis in Airplane Crashes: Creation of a Web-based Knowledge System ... 46
    Daniel I. Cheney, Safety Program at the Federal Aviation Administration
    United States

14. Integrated Disaster Research: Issues Around Data ... 49
    Jane E. Rovins, Integrated Research on Disaster Risk Program of ICSU
    China

15. Understanding Brazilian Biodiversity: Examples Where More Data Sharing Makes the Difference ... 54
    Vanderlei Canhos, Reference Center on Environmental Information (CRIA)
    Brazil

16. Social Statistics as One of the Instruments of Strategic Management of Sustainable Development Processes: Compelling Examples ... 58
    Victoria A. Bakhtina, International Finance Corporation
    United States

17. Remote Sensing and In Situ Measurements in the Global Earth Observation System of Systems ... 65
    Curtis Woodcock, Boston University
    United States

18. Discussion of Part Three by Symposium Participants ... 69

PART FOUR: THE LIMITS AND BARRIERS TO DATA SHARING ... 73

19. Data Sharing: Limits and Barriers and Initiatives to Overcome Them – An Introduction ... 74
    Roger Pfister, Swiss Academies of Arts and Sciences
    Switzerland

20. Consideration of Barriers to Data Sharing ... 78
    Elaine Collier, National Institutes of Health
    United States

21. Artificial Barriers to Data Sharing – Technical Aspects   81
   Donald R. Riley, University of Maryland
   United States

22. Scientific Management and Cultural Aspects   85
   David Carlson, University of Colorado
   United States

23. Political and Economic Barriers to Data Sharing: The African Perspective   89
   Tilahun Yilma, University of California, Davis
   United States

24. Discussion of Part Four by Symposium Participants   93

PART FIVE: HOW TO IMPROVE DATA ACCESS AND USE   97

25. Government Science Policy Makers' and Research Funders' Challenges to International Data Sharing: The Role of UNESCO   98
   Gretchen Kalonji, UNESCO
   France

26. International Scientific Organizations: Views and Examples   102
   Bengt Gustafsson, CFRS/ICSU
   Sweden

27. Improving Data Access and Use for Sustainable Development in the South   107
   Daniel Schaffer, Academy of Sciences for the Developing World
   Italy

28. How to Improve Data Access and Use: An Industry Perspective   112
   John Rumble, Information International Associates
   United States

29. Production and Access to Scientific Data in Africa: A Framework for Improving the Contribution of Research Institutions   115
   Hilary I. Inyang, African Continental University System Initiative
   University of North Carolina, Charlotte
   United States

30. The ICSU World Data System   118
   Yasuhiro Murayama, National Institute of Information and Communication Technology
   Japan

31. Libraries and Improving Data Access and Use in Developing Regions   120
   Stephen Griffin, National Science Foundation
   United States

32. Developing a Policy Framework to Open up the Rights to Access and Reuse   125
    Research Data for the Next Generation of Researchers
    *Haswira Nor Mohamad Hashim, Queensland University of Technology
    Australia*

33. Discussion of Part Five by Symposium Participants   143

APPENDIXES   **147**

A: Symposium Agenda   148
B: Biographies of Symposium Chairs and Presenters   151
C: CFRS Advisory Note   162

LIST OF FIGURE AND TABLES

FIGURE 3-1 Distribution of Approved Project Cost	10
FIGURE 3-2 Articles Downloaded	11
TABLE 3-1 Ph.D. Output in Pakistan	13
FIGURE 15-1 speciesLink Network Architecture	55
FIGURE 15-2 Access to CRIA's Online Systems in 2010	56
FIGURE 16-1 HDI Index Adjustments Due to Inequality	59
FIGURE 16-2 Six African Countries: Overall Satisfaction with Life	60
FIGURE 16-3 Multidimensional Poverty Index of Six African Countries	61
FIGURE 16-4 Six African Countries: Overall Satisfaction with Life	62
TABLE 17-1 Image Acquisitions by Country	65
FIGURE 17-1 Landsat Web-enabled Monthly Statistics	66
TABLE 19-1 Daily Newspapers per 1,000 People	75
TABLE 19-2 Television Receivers per 1,000 People	75
TABLE 19-3 Radio Receivers per 1,000 People	75
TABLE 19-4 Internet Growth 2000-2010	76
TABLE 21-1 2010 Internet World Statistics	81
FIGURE 21-1 World Internet Penetration Rates by Geographic Region in 2010	82
FIGURE 21-2 The International Reach of the Internet2 Network	83
FIGURE 21-3 Sample Bandwidth for African Universities	84
FIGURE 31-1 The World's Capacity to Store Information	121

FREQUENTLY USED ACRONYMS

BISO	Board on International Scientific Organizations
BRDI	Board on Research Data and Information
CERN	European Organization for Nuclear Research
CFRS	Committee for Freedom and Responsibility in the Conduct of Science
CODATA	Committee on Data for Science and Technology
GEOSS	Global Earth Observation System of Systems
HDI	Human Development Index
ICSU	International Council for Science
ICT	Information and Communication Technology
ICTP	International Center for Theoretical Physics
IPY	International Polar Year
IRDR	Integrated Research on Disaster Risk
NIMR	National Institute for Medical Research
NOAA	National Oceanic and Atmospheric Administration
NREN	National Research and Education Networks
OECD	Organisation for Economic Co-operation and Development
SDSS	Sloan Digital Sky Survey
TWAS	The Academy of Sciences for the Developing World
USGS	United States Geological Survey
WMO	World Meteorological Organization

1. Welcoming Remarks

Charles Vest
National Academy of Engineering, United States

Welcome to this International Symposium on the Case for International Sharing of Scientific Data, with a focus on developing countries. This symposium is one of over 200 activities organized each year by the U.S. National Academies. In the United States we have three academies: the National Academy of Sciences, the National Academy of Engineering, and the Institute of Medicine. These academies together operate the National Research Council (NRC). We are chartered by the U.S. Congress to provide objective advice on matters of science, technology, medicine, and health.

The theme of this international symposium is the promotion of greater sharing of scientific data for the benefit of research and broader development, particularly in the developing world. This is an extraordinarily important topic. Indeed, I have devoted much of my own career to matters related to the concept of openness. I had the opportunity to promote and help build the open courseware program at the Massachusetts Institute of Technology (MIT). This program has made the teaching materials for all 2,000 subjects taught at MIT available on the Web for anyone, anywhere, to use anytime at no cost. In countries where basic broadband was not available, we shipped it in on hard drives and compact disks. Its impact has been worldwide, but it has surely had the greatest impact on the developing world. I am also a trustee of a nonprofit organization named Ithaca that operates Journal Storage (JSTOR) and other entities that make scholarly information available at very low cost.

Even more to the point, however, is the fact that the culture of science has been international and open for centuries. Indeed, the scientific enterprise can only work when all information is open and accessible, because science works through critical analysis and replication of results.

In recent years, as some scientific data, and especially technological data, have increased in economic value frequently has caused us to be far less open with information than business and free enterprise require us to be. Indeed, the worldwide shift to what is known as open innovation is strengthening every day.

Finally, since the end of World War II, the realities of modern military conflict and now terrorism have led governments to restrict information through classification. This is important, but I believe that we classify far too much information. The last thing we need today, at the beginning of the twenty-first century, is further arbitrary limitations on the free flow of scientific information, whether by policies established by governments and businesses, or by lack of information infrastructure.

For all these reasons, the international sharing of scientific data is one of the topics of great interest here at the National Academies and has been the subject of many of our past reports. This is the primary reason why this symposium has been co-organized by two of our boards, both within the NRC's Policy and Global Affairs Division—the Board on International Scientific Organizations (BISO) and the Board on Research Data and Information (BRDI).

The purpose of BISO is to oversee and coordinate the work of more than 20 U.S. national committees corresponding to the International Council for Science (ICSU) and its different international scientific unions. It is the National Academies' lead on relations with the international scientific, engineering, and medical organizations. The Board chair is Dr. Cutberto Garza of Boston College, who is a member of our Institute of Medicine. One of the former members of BISO and a National Academy of Engineering member is Professor Farouk El-Baz of Boston University, the chair of this symposium. The director of BISO and the principal co-organizer of this symposium is Kathie Bailey Mathae.

The mission of BRDI is to improve the stewardship, policy, and use of digital data and information for science and broader society. It undertakes studies, workshops, symposia such as this one, and many other activities in pursuit of that change. The current chair of the Board is Professor Michael Lesk of Rutgers University, also a member of the National Academy of Engineering, and the vice chair is Dr. Roberta Balstad of Columbia University, who is on the steering committee of this symposium. Other members of the symposium steering committee include Professor Barbara Andrews of the University of Chile in Santiago; Dr. John Rumble of Information International Associates in Oak Ridge, Tennessee; Professor Tilahun Yilma of the University of California, Davis in Sacramento, California, and a member of the National Academy of Sciences; and Professor William Wulf of the University of Virginia and the former president of the National Academy of Engineering. Paul Uhlir, the director of the Board on Research Data and Information, is the other main co-organizer of this symposium.

Both boards are cooperating on this project with the Committee on Freedom and Responsibility in the Conduct of Science (CFRS), organized under the International Council for Science (ICSU). ICSU was founded in Paris in 1932 as a nongovernmental organization dedicated to strengthening international science for the benefit of society. The CFRS was founded in 1963 with a mandate to promote the ICSU principle of universality of science, which encompasses the freedom of movement, association, expression, and communication for scientists, as well as equitable access to data, information, and research materials. Professor Bengt Gustafsson of the University of Uppsala in Sweden is the chair of the CFRS, and Professor Roger Pfister of the Swiss Academies of Arts and Sciences is the executive director. Both are here today and will be actively participating in the symposium.

We all share the same world. We share its environment, its natural resources, and our common humanity. We must also share our knowledge. Addressing the great global challenges of sustainability, health, and prosperity are all well served by opening access to and sharing scientific and technological data and information.

**PART ONE:
SETTING THE STAGE**

2. Background and Purpose of the Symposium: Historical Perspective

Farouk El-Baz
Boston University, United States

To help people improve their lives, it is essential to contribute to the research processes and practices. Many examples exist demonstrating how the sharing of scientific data has improved people's lives and living standards. For example, these include better health and food safety data, but such improvements are not yet widely available in the developing world.

American and European researchers, and very recently researchers in the Middle East, have undertaken initiatives to promote data sharing with the developing world, for example, the sharing of remotely sensed environmental data. Also, there are similar examples from the United Nations organizations, such as the World Health Organization (WHO) and the World Meteorological Organization (WMO). These certainly have been very good developments, but they are insufficient.

I will give some examples from my country of origin, Egypt. In 1990 the Egyptian minister of agriculture asked, "We have heard about satellite images that can measure the area of land used for a specific purpose. Is this true?" He continued: "I have asked three agencies in Egypt a simple question: How much land in Egypt is under agriculture? When I received the results, however, these varied from 7.2 million acres to 5.5 million acres. How am I going to plan if I do not know whether it is 5.5 million or 7.2 million acres?" So, my team superimposed the satellite images of the same area in 1972 and in 1990, and showed the minister the amount of land that was used for agriculture on both dates. We also discovered something that was even more important: how much of that land has been transferred into urban areas in the 18 years since 1972 (i.e., the encroachment of urban areas over agriculture), which is even more dangerous. The Ministry of Agriculture began using that data. They trained their own people, and my team worked with them on proper procedures.

Then in 2009, this group of Egyptian researchers who knew how to use the data raised a warning to the Egyptian government. They said that in the past 20 years, the average loss of fertile land to urban growth in Egypt equaled 30,000 acres per year, which is obviously a very large number. If this were to continue unabated, then the agricultural land in Egypt would disappear in 183 years. The government then took very dedicated steps to change this situation. The researchers were able to raise the warning because of data and knowledge sharing. The images they used in this analysis were from the series of Landsat satellites that were started by the U.S. National Aeronautics and Space Administration (NASA) in 1972, and those satellites continue to produce images with even better resolutions to this day.

There is another critical point here, however: the data were shared freely and the Egyptian researchers were able to continue this work by themselves. This benefit was due to the fact that the scientific community worked with NASA (and more recently, the U.S. Geological Survey) for years. In 2008, NASA agreed that this satellite image data would be freely available, and from the time the use of the Landsat data has increased a hundred-fold.[1]

The barriers to sharing of data are many and, in my experience, can be stronger in developing countries. One barrier is a national attitude of protectionism. Many researchers in less-developed countries will say, "We have worked on this data. This is information about us. Why would we make it available to people everywhere?" A second barrier is the cost. Those same researchers would say, "We spent a lot of money on research and development on this system. After spending all of this money, why should we share it with others?" A third barrier is national security. There are many types of data behind a secure door,

---

[1] For a presentation about the Landsat data use, see Chapter 17 by Curtis Woodcock.

because the data appear to be related to national security. Yes, there are certain things that should be classified, but these should be limited and should include only the data that really relate to national security rather than just based on the indiscriminate perception of their importance to national security.

When data have been developed well, and then made available, this can make a huge difference in the lives of people or the economy of a developing country. For example, radar images from space were developed by NASA's Jet Propulsion Laboratory, and the radar imager flew on the space shuttle for the first time in November 1981. My team applied the new technology to locating ancient water resources in Egypt. The images that were brought back confirmed that radar waves could penetrate through desert sand, because it is fine grained and dry. Because of that, these images gave us a view of the land surface beneath the sand. That was the first time we could actually prove that there were channels of former rivers that are now dry and covered by sand. The radar gave us a map of all of those channels. When my team drew the maps of the channels, we knew that water was moving from one elevation to a lower one. We assumed that we would find groundwater sites, because water collected there in the past. Some of that water would evaporate and some of it would seep through the rock and be locked up as groundwater. That was our speculation, and it made sense to us, geologically and topographically.

Based on these findings, my research team picked a place in the southern part of Egypt and began to talk to the people there, including the Ministry of Agriculture. It took me 13 years to convince the minister to test drill. He finally approved two wells, and there was a great deal of clean water from around 20,000 years ago, where ancient rivers flowed over a sandstone substrate. Sandstone does not have much salt; the water is cleaner and sweeter than that of the Nile River. It comes out of the ground cool and clean, as if it has been refrigerated. As a result of that discovery, there are now 200,000 acres of land that are viable for agriculture. In that location they have now drilled over 1,000 wells, and are producing much of the wheat that is used for bread making in southern Egypt.

This is an example of something that happened when a group of scientists shared and studied data, and it has made a huge difference in the economy and knowledge base of Egypt. I also published an article on a proposed development corridor that is parallel to the Nile River. The data, including topographic maps, geological data, and space images became readily available on the Internet, which made it easy for others to utilize them. Egyptian geologists and geographers began to pick pieces of my proposal to use as research topics.

Similarly, data from NASA's Shuttle Radar Topography Mission (SRTM) provide very good feedback about the topography of an area. Using topographic data, researchers can see all kinds of places, including former rivers, channels, and more of the potential lakes where water collected in the past. My team did that for the area of Darfur in Sudan. We shared the data with the Ministry of Water of Sudan and the people in Darfur. This information is truly important, because the disaster in Darfur started with conflict over scarce water resources. This new information gave hope to the people of Darfur that they could have more water. Some of the places where there were no wells now have the potential of plentiful water.

In another example, the U.S. government, through the National Academies and the Civilian Research and Development Foundation (CRDF Global), made many scientific journals freely available to all universities in Iraq, and in 2009 the Iraqi government took over the funding responsibility. Computers and other infrastructure have been provided, so that the journals can be easily accessed. After the project began, the publication rate of Iraqi scientists increased significantly, based simply upon the fact that the researchers were able to read the literature and see the data that were available to them. As a result, they were able to figure out what they could contribute. The increase in the publication rate was immediately visible and recognizable.

In my view, therefore, scientists ought to continue to campaign and lobby governments to make more data freely and easily available, and only classify documents that raise legitimate national security concerns. As the three examples that I have described illustrate, the sharing of scientific data can produce a great deal of good, particularly in the developing world.

3. Why Is International Scientific Data Sharing Important?

Atta-ur-Rahman
UNESCO Science Laureate, Pakistan[2]

It is a great privilege and an honor for me to be talking to you today. I will be speaking about the world in which we live: the digital age, an age where truth is often stranger than fiction. I will also talk about the knowledge explosion and the new challenges and opportunities that this is offering. I will try to address how countries in the developing world need to transition from largely low-value-added agricultural economies to knowledge economies. A key factor is how they work together, how they share data, and how they collaborate on a regime of openness that crosses geographical boundaries. Then I will focus on Pakistan and what we have been able to accomplish in these areas.

We live at a time in which distances have become much less meaningful. The opportunities of networking have created huge new vistas for cooperation. Information and communication technologies (ICTs) have been driving growth, whether it is commerce, telemedicine, governance, or geographic information system (GIS) mapping. I am a professor of organic chemistry, and I have 12 superconducting nuclear magnetic resonance spectrometers in my center, from where I am speaking now. These can be controlled remotely. Sometimes, when we need to fine-tune them, we no longer we need to come here; we can operate them from a distance.

The blind are today able to see with their tongues. How do they do that? They have a little device on their glasses, and this takes the optical signals, which are then converted into electrical signals by a device in their pocket. These electrical signals are then transmitted to a lollipop-like device in the tongue, and this then sends the signals to the brain. They can distinguish between a knife and a fork, and distinguish lift buttons. We now have paralyzed people able to move wheelchairs with thought control. They wear a skull cap, and then encephalographic signals and blood pressure changes are sensed. They think about, for example, moving their wheelchair to another location, and it does that. Stem cells are transforming the way medicine is going to be practiced in the future. I could go on and on. All this is connected to research, and research is connected to data and the free availability of data.

In the world that we live in now, the key requirements for socioeconomic development are as follows:
- Human capital, with the necessary knowledge and skills.
- Technology.
- A society where innovation and entrepreneurship can flourish.
- Mechanisms for all this to be put into place, through data sharing, knowledge generation, and application.

ICTs are reducing the gap between the richer and poorer countries. The ICTs are offering opportunities to developing countries that did not exist before.

What are the challenges? Many of the advanced countries have diminishing numbers of young people who are opting for careers in science or in education. In the developing countries, however, we have a different set of issues, starting with the serious lack of vision among the leadership—they do not realize the importance of knowledge in the process of socioeconomic development.

In a world where innovation drives progress, it has become easier to catch up, given the political will. We have the case of China, for instance. Never before in the history of humankind have so many people's

---

[2] Presentation given by videoconference.

lives been transformed so quickly—a billion-plus—as has happened in China. This is an example at the country level. At an institutional level, the Massachusetts Institute of Technology (MIT) highlights what one high-quality institution can accomplish. From MIT alone, some 4,000 companies have sprung up, with annual sales of $240 billion and employment of more than 1 million people. All these companies would constitute the 18th largest economy in the world[3].

What are the constraints, from the point of view of a developing country, on using and sharing knowledge, and progressing toward knowledge economies? As I said before, it is mainly the lack of vision, strategy, and an action plan on how to transition to knowledge economies. The efforts of most developing countries are diffused, as the countries attempt to do too little in too many areas. The result is that the impact is not there. Also, they are unable to use the data that may be available because they lack:

- A critical mass of quality researchers.
- A robust ICT infrastructure.
- Adequate funding for research and libraries.
- Compelling incentives for academics to publish.

We also find that there is a new wave, which is fairly big and growing, of open-access journals. Governments, institutions, or authors can either subsidize the journals or pay in advance and get their articles published and be viewed by millions, instead of a few thousand or even just a few hundred who may subscribe to those journals. Similarly, there are open-access books—complete books or chapters. Even lectures are becoming open access, as with MIT, as described by Dr. Vest in Chapter 1. Of course, we have the issue of quality, as in all publications. It is important that we publish only in those journals that are properly peer reviewed, so that quality is not compromised.

Take again the example of the MIT open-course initiative. I think this is an excellent initiative from which the Pakistanis have benefited considerably. We created the MIT mirror Web site within Pakistan, and all of MIT's open-access materials became quickly downloadable to all students in the country. I wish that other universities would follow suit and make some of their materials available. Apple iTunes U is another excellent initiative, where we can go to the Web sites of many universities in Europe or the United States and listen to some excellent lectures. There are many other similar initiatives, and all of these are then providing new opportunities.

We live in a world where the only constant is change. The challenge is change management. Do we guide this change, or do we get buried under it?

Let me now talk about Pakistan. I was appointed as the minister for science and technology and, later, the federal minister responsible for higher education. For about 9 years, from 2000 to 2008, I was involved in trying to boost science and technology and higher education in Pakistan.

I managed to convince the government that if they are serious about progress and about transitioning to a knowledge economy, then they have to invest. We persuaded the government to give us a 6,000 percent increase in the budget for science and technology and a 2,400 percent increase in the budget of higher education. I know it sounds too good to be true, but it is. This allowed us to undertake a variety of different reforms, in which access to data and its use was critically important in all that we did. The first step was to prepare a national strategy and action plan. Where are we going? In 20 years, will we be building ships, specializing in computer chips, or a world-class manufacturer of pharmaceuticals? What is the roadmap for development—a national vision, strategy, and action plan?

The cabinet asked me to prepare a comprehensive plan to transition to a knowledge economy. Then we

---

[3] Available at http://web.mit.edu/newsoffice/founders/Founders2.pdf

looked at different sectors and specific programs and projects that would be conducted in all these sectors, and came forward with a very clear strategy. I had 12 colleagues with Ph.D.s in economics to help me establish priorities. We scientists often think that our own subject is the most important in the world; for a reality check, we got external inputs. As a result, we set up a number of committees. There were interactions with thousands of scientists and engineers and economists within the country and across the world—our own diaspora, other specialists, government planners, people in different ministries, people in the provincial and federal governments, and people involved in different organizations. After doing a strengths, weaknesses, opportunities, and threats analysis and identifying the specific strategy, we came forward with a very clear and crisp action plan.

One of the instructions that I had given was that I do not want a long wish list, because the last thing that we want is to have long lists of things that cannot be done. I said, just prioritize and give me half a dozen or, at the most, a dozen top priorities in each sector that will have maximum impact on the process of socioeconomic development. This involved identification of the technology gap, the policy gap, the innovation gap, and then a strategy for how to fill those gaps. These were the key questions we tried to answer:

- What specifically is to be done in each area?
- Who are going to be the players?
- In what time frame?
- What are the human resource requirements?
- What are the costs?
- What is the international rate of return and the impact on the national economy?

The 300-page plan was approved by the cabinet a few years later. This then linked up everything to a national strategic action plan.

The second constraint for using data and for benefiting from it is the lack of a critical mass of high-quality researchers, the people who are going to be using this data and employing it for publications or for a variety of different purposes. For this, we had to invest massively in the process of human resource development, so that we would develop the national abilities in different fields.

In Pakistan, we have a population of about 160 million, of which 85 million are below the age of 19 years. Here is a country where 54 percent of the population is a young population. This presents both a challenge and an opportunity. How do we educate these large numbers of young people? How do we provide schools, colleges, and universities, as well as jobs, after they graduate? It is a huge challenge. It is an opportunity also, because if you tap into this pool, then something wonderful can happen.

But how do we utilize this pool of people? First, we have to excite young minds about the wonders of science. In fact, I write articles every Sunday in our leading national newspaper, titled, "The Wondrous World of Science." I have done that for the last 2 years, where I am basically saying, "Hey, science is fun. It is not about a 9-to-5 job. You are going to be paid to have a love affair with science, if I may put it that way. You are really going to enjoy life, because it is exciting. There are so many exciting things that are happening around it. Just feel the buzz and be a part of this buzz." We have to excite young minds about the wonders of science.

We then have to select the brightest. In this pool of 85 million, surely there are hundreds of thousands of geniuses. Can we pick them out and polish them and then train them at top institutions abroad and let the light of knowledge then pour from them? For that to happen, we have to attract them back from abroad, not by legal bonds, but by creating an enabling environment, which involves good salaries; access to research funding, literature, and instrumentation; and the creation of clusters. Just having a certain number of colleagues with whom they can talk and discuss their problems is important. If you are working in biotechnology or in information technology or in some other area, it is important to have a clustering of good people.

Perhaps the most dramatic step we took was to persuade the government to completely change the salary structures and make the salaries of professors in universities five times the salaries of federal ministers in the government. We introduced a new tenure-track system; however, we linked this to performance. There is no point in paying weak people high salaries. That would be just a waste of money. We told them that they will be on contract for 3 years, and then they will be renewed for another 3 years, after initial assessment by peers. They have to be evaluated again after 6 years, by an international panel of experts before their position is made permanent. The tax structure was also reduced from 35 percent to 5 percent.

We then started identifying the brightest through a national examination, and sending many of those abroad. Nearly 11,000 fellowships were awarded during that time. The world's largest Fulbright Scholar Program was initiated, half funded by our government and half by the U.S. Agency for International Development. Each returning scholar was offered a grant of up to $100,000. When they return, however, they often come back to barren environments. It is critically important that we give them a red-carpet welcome and provide them with the opportunities to start productive work as fast as possible. They could apply for a research grant, but one condition was that there should be a foreign collaborating academic as a Co-Principal Investigator. In fact, they can apply for a research grant a year before they are to return. This has also led to many linkages with their own universities. They were guaranteed jobs in advance: A year before they returned, they had jobs to come back to. It was all very well thought out. It was a billion dollar program of foreign scholarships in various fields of science and engineering, social sciences, et cetera.

Figure 3-1 shows how the money was spent: The largest part of the distribution was for human resource development, and the rest was for infrastructure, research, and access to instrumentation.

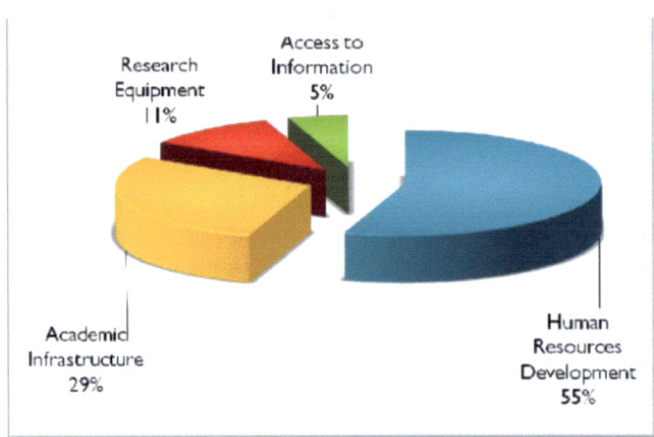

FIGURE 3-1 Distribution of approved project cost.
SOURCE: From the speaker's presentation at the symposium.

First, we invested in the development of human resources, but then we needed to build the knowledge highways. We need to have the infrastructure on which data can be shared and data can travel. These are in many ways far more important than the motor highways we are used to. I was also the minister for information technology and telecommunications. We did a number of things to widen these highways and to make them efficient. For instance, we placed a satellite in space, Pakistan's educational satellite, PAKSAT-1, which has footprints over parts of Africa and Asia, and the Middle East. We also established a virtual university with its own recording studios and used our satellite for communication purposes. Moreover, we rapidly spread the Internet. In 2000, only 29 cities in Pakistan had access to the Internet. This increased to 2,000 towns, cities, and villages that had Internet facilities in 2003. Fiber optic was rapidly expanded from 40 cities to 1,000 cities between 2000 and 2005. Internet usage exploded and continues to do so. The costs of bandwidth were brought down from $87.00 U.S. dollars per month for a 2 Mb line to only $900 per month, and they are among the cheapest in the world today.

Mobile telephony has also expanded. From 300,000 mobile phones, the number suddenly exploded to 110 million-plus mobile phones, as a result of some of the steps that we took. This became the foundation for the establishment of the Pakistan Education and Research Network, which connected all universities together with high-speed Internet access. That has been further expanded with a 1-gigabit connectivity to every university, which is then connected to 10-gigabit loops around major cities.

A very robust system of knowledge highways developed. The next constraint was the question of content and useful, meaningful data that researchers can use. In response, we established the digital library.[4] If you walked into a library 10 years ago in Pakistan, you would have found perhaps half a dozen of the latest international journals. Today, as shown in Figure 3-2, every university in the public sector and most universities in the private sector have free access to 25,000 journals and to some 60,000 textbooks from more than 220 international publishers. These are keyword searchable and downloadable. When I say free, I mean that the government, and specifically, the higher education ministry, pays for this vast repository of knowledge. They are also available to students, not just from the universities, but also from their homes, so that if they are working until late in the evening, they can access these databases.

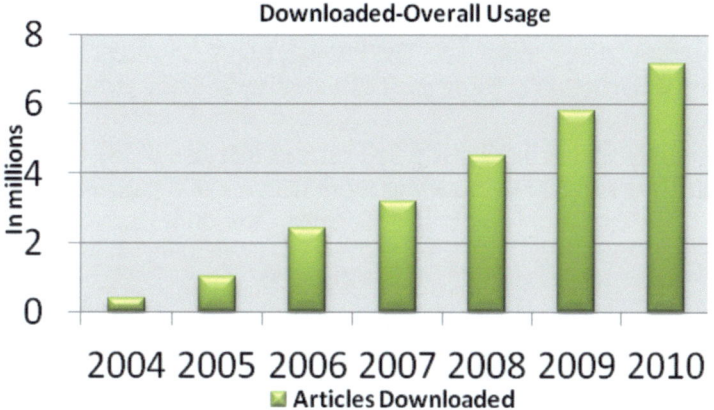

FIGURE 3-2 Articles downloaded
SOURCE:: *From the speaker's presentation at the symposium.*
Credit: Higher Education Commission for Pakistan's National Digital Library Program

There are 75,000 e-contents, whether it is Springer or McGraw-Hill or the ISI Web of Science. We also introduced a search engine, SciFinder Scholar. We worked with the University of Lund in Sweden to develop a one-window search engine so that users do not have to go to different Web sites; they can just search all these journals or books through one window. That has worked very well for the last 6 or 7 years. The usage statistics grew very quickly, and they continue to increase.

One excellent teleconferencing facility that we built is the one that I am speaking from now. Such facilities are available in every public-sector university. Here in Karachi University, we have five such

---

[4] Available at www.digitallibrary.edu.pk.

facilities. We started a series of lectures and courses, which are developed by top professors from across the world and are benefiting other developing countries as well, such as Sri Lanka, Thailand, and other countries. They are delivered through these videoconferencing facilities. The professors then supervise the exams, which are held by the local partner in the department. They also supervise the marking of papers. Then the universities recognize these as part of their programs' credit requirements. This is allowing us to link up both with the best of the best in the West and with our own top people in Pakistan, and share faculty and resources with one another.

Pakistan's telemedicine program was also based on an initiative that I had started in 2001, when we sent selected doctors to the United States. When a massive earthquake occurred in the north of the country, the telemedicine program proved invaluable, because we could deploy remote services to the affected regions in a timely fashion. More than 4,000 teleconsultations have been held and this continues to grow. These repositories of knowledge (i.e., the databases and the literature) are all freely available in Pakistan.

One problem that we had to address was controlling the improper use of data. With access to data and publications growing, the menace of plagiarism has to be controlled. We set up a national plagiarism policy and distributed the required software. This is available in all universities and research institutions, and all theses and research papers are carefully checked before the universities process them.

What has been the impact of all this provision of data, this development of human resources that share the data, and this access to knowledge? Between 2003 and 2009, we had a tripling of students enrolled in higher education. This continues to grow, but is still very low. Only about 5.2 percent of our youth between the ages of 17 and 23 have access to higher education. It should be about 15 or 20 percent. In South Korea it is almost 90 percent. The enrollment has to grow in Pakistan, but there are challenges. The number was well below 2 percent when I took over. The number of public-sector universities also grew from 59 in 2000 to 127 universities in 2009. There are 132 today, in both the public and private sectors.

One of the most remarkable things is that there was a 600 percent increase in ISI-abstracted publications in the 5-year period between 2004 and 2009. There was more than a 1,000 percent increase in citations—after removing self-citations—in The Thomson Reuters Science Citation Index[5]. There also was fast growth of papers published.

As shown in Table 3-1, the Ph.D. output also grew phenomenally, although it still is very low for a country the size of Pakistan. In the 55 years between 1947 and 2002, Pakistan produced about 3,200 Ph.D. degrees, and in the subsequent 7 years, about as many Ph.D. degrees were produced, a little more than 3,000.

---

[5] Available at http://thomsonreuters.com/products_services/science/science_products/a-z/science_citation_index/.

TABLE 3-1 Ph.D. Output in Pakistan

Discipline	1947 – 2002 [55 years]	2003 – 2009 [7 years]
	Number	Number
Agriculture & Veterinary Sciences	363	450
Biological & Medical Sciences	589	601
Engineering & Technology	14	131
Business Education	11	58
Physical Sciences	688	677
Social Sciences	899	739
Arts & Humanities	663	377
Honorary	54	4
Total	**3,281**	**3,037**

SOURCE: From the speaker's presentation at the symposium. Credit: Higher Education Commission for Pakistan's National Digital Library

However, numbers without quality can do more damage than good. We introduced a system in which all doctoral theses had to be reviewed by at least two or three experts in technologically advanced countries, and only when there was unanimity of opinion that the work was of high quality could the thesis be processed further. We now have several universities in Pakistan that are among the top 300 in the world, including the National University of Science and Technology.

Nature, one of the world's leading science journals, has published a number of editorials on what Pakistan did. The first of these was on November 27, 2007, which was titled "The Paradox of Pakistan." In this strange country, which is torn by terrorism and bombings and all sorts of difficulties, there also is a wonderful program in higher education, science, and technology. On August 28, 2008, when President Musharraf had left, the editorial was entitled "After Musharraf." It said that we must not go back to the same Stone Age that existed previously. Something exciting has happened in Pakistan, and it must be continued.

The last *Nature* article was on September 22, 2010, which talked about what we had done. I was humbled and somewhat amused when it called me "a force of nature." I also won the highest civil award from the government of Austria for these changes and the TWAS Prize, which was given to me for institution building. The Royal Society of London, of which I am a fellow, has come forward with a book on what is happening in the Islamic world. It is entitled *A New Golden Age*. It talks about Pakistan and says that it is the best practice model to be followed by other developing countries.

The future of our country lies in our youth. It is only through the unleashing of their creative potential that we can hope to move forward. We picked 12 students who appeared to have high aptitudes in local high school examinations, and sent them to international Olympiads in mathematics, physics, chemistry, and biology. One of them came back with a silver medal and two came back with bronze medals. This was only after a couple of months of preparation. There is a huge potential that we have in countries such as Pakistan, but it will only be possible to tap into it if we realize that our real wealth lies in our children.

When I came to Pakistan after 9 years at Cambridge University, I dreamed that I would try to set up a center in Pakistan that would be so good that students from the West would come to study here. In my center here in Karachi, we have 350 students doing Ph.D.s in organic chemistry, biochemistry, pharmacology, and other related areas. In the last 5 years, we have had 120 German students who have come to study chemistry and science.

4. DISCUSSION BY THE SYMPOSIUM PARTICIPANTS

PARTICIPANT: What kind of planning is Pakistan doing for the next decade?

DR. ATTA-UR-RAHMAN: In my speech, I mentioned that I was asked by the government to prepare a national roadmap. We prepared a 15-year roadmap for Pakistan. The plan was divided into three 5-year periods, for planning purposes, so that we can generate employment, transition to a higher value-added economy, and use our most important resource, our youth, for future development. We have had some serious problems in the last two and a half years, with the new government facing many financial constraints and freezing some of these programs, but the new education policy, which was approved by the present government, promises a fourfold increase in the budget for higher education and science-related disciplines. If that materializes, I hope that things will start changing again rapidly.

PARTICIPANT: One of the things that seismologists hear from their colleagues in Africa and South America is that the mineral exploration industry often hires their best students, who then do not go into academia. That is quite surprising to us in the United States, because here it is regarded as a good synergy that industry and academia work together, and it contributes to the strength of academic programs. I am wondering what is your view about this sort of back-and-forth of students going into industry?

DR. ATTA-UR-RAHMAN: The problem in Pakistan and many other developing countries is that we do not have strong private-sector research and development partnerships. Most of our institutions do not have research facilities. What worries us is not students being taken away by local industry. I totally agree that we need to have synergy and interaction, but they are being taken abroad, and we are losing a valuable treasure that we have. For that, we have to create the enabling structure that I talked about—the research facilities, the salaries, and so on. They do not have to work in any government institution, as long as they come back and work in Pakistan. It could be in the private sector. So our problem is a different one. For innovation to flourish in a country like Pakistan and other developing countries, you need to have a number of things in place. This includes ease of doing business, access to venture capital, intellectual property right regimes, presence of technology parks, legal infrastructure, and the provision of greater incentives for private-sector research and development so that innovation can take off. All these measures have to come together before the development, innovation, and entrepreneurship can take place. That is still largely missing in Pakistan and in most other developing countries.

PARTICIPANT: I would like to address one of the points you made regarding the volume of publications emerging from the developing world as a share of the world total. You mentioned that it had risen from, I think, 21 percent to 32 percent. This assumes that the actual database of publications that is being counted is a constant, and it is not. Since 2005, the Thomson Reuters Web of Knowledge has quite dramatically changed the volume of publications that it accesses from developing countries, which is a very important change. For example, the number of South African journals that are indexed has gone up nearly threefold. The same is true of Brazil. I took the liberty of looking at the numbers for Pakistan during your talk; there were only two Pakistani journals on the Science Citation Index in 2005. That rose to 14 in 2010. So without detracting from the achievements of ourselves in the developing world, we need to be sure that we benchmark correctly.

DR. ATTA-UR-RAHMAN: What I was talking about was not the number of scientific journals that are published from these countries. The two journals that you are talking about are the two journals from

Pakistan that are included in ISI Web of Science and have an impact factor. There are, I think, about four or five journals. I was talking about international publications in international journals from the developing world.

The data that you talked about also needs to be tweaked a little, because although there is an increase, this increase is largely because China is counted as a developing country. If you take away China, then I think you will not find that there would be much of a change in the rest of the developing world. This data was taken from the United Nations Educational, Scientific and Cultural Organization (UNESCO) 2010 *World Science Report*, which has grouped the developing countries together, including China.

PARTICIPANT: There has been some news recently about the Pakistani government's interest in devolving some of the responsibility from the central government to the provinces for higher education and other research activities. Could you comment on that?

DR. ATTA-UR-RAHMAN: This has been an impending disaster in Pakistan in the last few weeks. What has happened is that because the Higher Education Commission was an honest and merit-based organization, it would not bow to political pressure. The Commission found that about 50 of our parliamentarians had forged degrees, and it reported it. Another 250 parliamentarians' degrees were in the process of being investigated. Certain young people in the government then decided to cut this organization to pieces. I took a very strong stand and wrote a number of articles in national newspapers. I went to the Supreme Court last week, and the decision from the Supreme Court of Pakistan was that the Higher Education Commission is protected under the constitution. This is where it stands now. I understand that there are reverberations. The government is thinking of changing the law, so there may be more to come.

PARTICIPANT: This is one illustration of the fact that if governments are not working for the good of the people, they can create all kinds of problems. We have to watch for that, but things are changing drastically. I just came back from Egypt. I have been talking to groups of young revolutionaries. Indeed, their outlook is very positive. They know exactly what the corruption within the government was like and what the results were. They are looking forward to new ways of doing things, both with the government and by themselves. The one thing they asked me to convey to the world is that they are ready and willing to work with anybody, especially in science and technology. They need help and they are willing to work harder than anyone else.

**PART TWO:
STATUS OF ACCESS TO SCIENTIFIC DATA**

5. Overview of Scientific Data Policies

Roberta Balstad
Columbia University, United States

We have heard a lot about the practical economic and applied implications of having open access to data, but we should not lose sight of the benefits to science. One of the reasons that access to data are becoming so important is, of course, that the technology has changed, and that we can deal with massive databases in a way that we could not have dealt with them 20 or 30 years ago. Another reason is in the very nature of the scientific process itself. What is science? For many people, it is simply experimentation and testing. That narrow definition has been modified in recent years to include experimentation, observation, and testing. For other people, science is really a matter of modeling and projections. If you cannot project something accurately, many believe, it is not science. So you need data for projections, too.

Equally important, scientific research is increasingly evolving into "data-intensive science." You read about it in the field of health care, for example, where scientists combine data from 20, 30, or 100 different studies to get a larger base in order to analyze and investigate topics that are impossible to pursue in a small, intensive study of perhaps 20 individuals. This is also true in a number of other fields. Data-intensive science relies on open access to data from all sectors, because only then are scientists able to combine datasets to ask new types of questions.

Scientists are able to address much broader questions in data-intensive science than they could if they were responsible for collecting their own data for every study that they conduct. Increasingly, for example, we find that governments collect much of the scientific data that we use. These databases in many countries are open. We would like to see them become more open in even more countries so that scientists can use them.

Open access to data advances science. It improves descriptive, comparative, and observational science; it enriches modeling and prediction; and it makes it easier to test and retest propositions using the same databases. That, of course, goes back to the philosopher of science Karl Popper, who said that true science is science that can be tested, that is falsifiable, and that you can prove wrong. To do that, you have to have access to data.

A second reason for providing better access to scientific data, in addition to advancing science, is that it levels the playing field for scientists from smaller or less-developed countries so that they are able to conduct data-intensive science using publicly available data. In short, data access makes a principal resource of scientific research available to all.

Traditionally, data access policies were quite restrictive in terms of both policies and practices. Data were held to be the private property of a scientist. At the end of doing a dissertation, we had a body of data that we could mine for a long time. That was considered to be the property of the scientist and that was what made his or her work significant. In other cases, the kinds of data that Professor Farouk El-Baz was talking about (e.g., remote-sensing data) were often seen as a national asset that had to be protected.

Data were also seen as a commodity that had economic value for the scientist or, more often, for the government that sponsored the data collection. When science becomes a commodity, obviously, those who collect data begin to think about marketing the data, and then they easily slide into charging for data in a for-profit or even not-for-profit setting.

To summarize, the benefits of changing from restricted to open data access policies are as follows:

- Open access to science contributes to innovation and economic growth.
- Scientific advances, both substantive and methodological, are now data intensive and require open access to scientific data.
- The cost of research is reduced. This is very important right now in most countries, because there is often less money available for research. To keep science alive and vital, open access to data is a real advantage.

Limiting access to data—the other side of that coin—results in higher research costs, lost opportunities, barriers to innovation, less less-effective scientific cooperation, suboptimal quality of the data (since no one is working with them and cannot provide corrections to them), and a widening gap between the Organization for Economic Co-operation and Development (OECD) countries and the developing countries.

In pushing for open access to data, however, we must acknowledge that there are some legitimate reasons for limiting access to public data:

- National security and public safety.
- Personal privacy and confidentiality, which are protected in many countries.
- Proprietary rights of private-sector parties. No one is talking about forcing open access on research that a company has done in order to advance its product.

Internationally, there have been a number of activities that have advanced open access over the past 50 or 60 years. A big impetus to open access to data was the International Geophysical Year (IGY) in the 1950s, a massive global-scale data-collection effort that stressed open access to the scientific data collected under the aegis of the IGY. One of the results of the IGY was that the International Council for Science (ICSU) formed the World Data Centers. In order to become a World Data Center, a center had to agree that it would provide scientific data to whoever asked. That does not seem to be required anymore, but it was at the time, particularly because a major goal was to make sure that data were available both to scientists in the West and scientists in the Soviet bloc. The Iron Curtain divided scientists as well as politicians, and the World Data Centers were meant to overcome the limits to exchange of data among scientists.

When the Group on Earth Observations formed the Global Earth Observation System of Systems (GEOSS) in 2005, it established the following open data principles:

- There will be full and open exchange of data, metadata, and products shared within GEOSS, recognizing relevant international instruments and national policies and legislation.
- All shared data, metadata, and products will be made available with minimum time delay and at minimum cost.
- All shared data, metadata, and products being free of charge or no more than cost of reproduction will be encouraged for research and education.

In 2007, the OECD made a strong stand on behalf of open data access, recommending that data policies show openness, flexibility, transparency, legal conformity with existing laws, protection of intellectual property, formal responsibility for the data, professionalism, interoperability, data quality, data security, data efficiency, accountability, and sustainability.

There has been gradual movement toward even more openness in data in the United States as well. In the 1980s, the Reagan administration proposed a policy of commercialization of all data that were collected under grants supported by the National Science Foundation (NSF). That would have meant that investigators would have to sell their data to anyone who wanted to use it. One division in the NSF, the Division of Social and Economic Science, established a policy of not making a grant to anyone who would not agree to put their data in a publicly accessible data archive before receiving the grant. If the grantees did not follow through with what they had proposed, they would not receive another grant. The policy had limited success because it only covered one division, and it went counter to the national policy. However, in 1991, the U.S. Global Change Research Program, backed by the White Houce Office of Science and Technology Policy (OSTP) established a policy of open access on data related to global change. In 2005, the National Institutes of Health required data management plans for all of its large grants. And this year the National Science Foundation promulgated a new data management policy, which requires all grant proposals to include a data management plan. In sum, the role of data—and policies for data—are changing rapidly.

6. Implementing a Research Data Access Policy in South Africa

Michael Kahn
University of Stellenbosch, South Africa

Today I am going to speak about three things, broadly: South Africa as the gateway for the BRICS Group, whose members are Brazil, Russia, India, China, and South Africa; its innovation system and policy; and prospects for research data policy.

It is safe to say that South Africa represents, by world standards, a relatively small innovation system that is quite dynamic. In the apartheid years, it engineered all sorts of bad things, but also some good things, especially in the fields of health, plant and animal science, ecology, and environment. Those deep skills prevail into the present. That is essentially why South Africa can play an important role of being a higher-education hub for the rest of Africa.

The story in Africa is very interesting. There are many African countries that are among the fastest-growing economies in the world. This morning CNN referred to China as growing at a mere 9.7 percent per year at the moment. The country that comes immediately after that is Ethiopia, at 8.5 percent. Angola, Chad, Democratic Republic of Congo, Ethiopia, Mozambique, and Zambia will all have growth projected over the next 5 years well in excess of 7 percent.

South Africa is currently growing around 3 percent and is struggling to break out of what appears to be a natural confine of around 4 percent. But I want to draw your attention to scientific production in South Africa. If we take the country's scientific article production, as recorded on Thomson Reuters Institute for Scientific Information (ISI), South Africa is the clear leader in Africa. The real question is whether growth can be driven by science and research and development (R&D), or whether growth is driven by industrialization.

Among the universities in Africa, the top 10 are all South African—hence, the higher-education hub. Also, there are a number of fields in which South Africa has scientific impact somewhat above the world average, such as immunology and space science. There is also a high level of particular expertise and activity, and therefore, necessarily, a great volume of research data in agriculture, environment, ecology, and geosciences. Furthermore, I should mention the increasing number of domestic journals that are indexed to Thomson Reuters ISI.

When talking about data policy, you can only talk about it in the context of your entire innovation system. In the case of Pakistan, mentioned this morning, the private sector plays a very small role. In the case of South Africa, the private sector is the largest performer of R&D, but also the smallest producer of scientific output in the form of articles, unlike Japan and even the United States, where many articles emanate from private-sector addresses.

When you talk about an innovation system, you talk about the main actors: business, higher education, and government. These operate in the pursuit of innovation activity, of which R&D is but one activity. You also have to look at your financial system; your cultural and political norms; the regulatory framework; the legal framework, including intellectual property; and information policy. If any one of these is suboptimal, the others are not going to flourish.

In South Africa, we have the National Research and Development Strategy of 2002, which brought about some reorganization of the science system related to reporting lines, which may or may not be significant. It did lead to the development of changes to intellectual property law, as well as the introduction of a forward-looking incentive for R&D, which rewards R&D, much as happens in the United States, with a

150 percent deduction in taxation. It also led to some initiatives in human-resource development, such as a research chairs project modeled on that of Canada, and the introduction of centers of excellence in the universities in various fields.

This has been followed by an innovation plan in 2008, which plays into the same theme of grand challenges that you find in many countries. The five grand challenges identified are all highly data-intensive areas:
1. Space science (i.e., remote monitoring and telemetry).
2. Energy, the hydrogen economy, and new materials or catalysis.
3. Farming to pharmaceuticals and biotechnology, plant and animal science.
4. Global dynamics (i.e., climate change) and remote sensing.
5. Human and social dynamics (i.e., social sciences).

Then we have the introduction of the Technology Innovation Agency in 2010, with a mandate of converting R&D findings into commercial prospects. These initiatives, of course, operate within the larger context of other national laws on employment equity as well as immigration law. These various thrusts are now being driven forward through an industrial policy. South Africa is a very important gateway in Africa, through its work with the New Partnership for Africa's Development. We have also been active in promoting international networking through our centers of excellence in high-speed computing.

What about the prospects for research data policy? The country has a science system—arguably going back a century and a half—to the onset of large-scale mineral exploitation, so it is not a newcomer. There are many countries in Africa where science systems date back perhaps 50 years or less. We have been at it for a long time, in the same way that Brazil has been at it for a very long time.

South Africa is data rich at the system level. We have well-quantified data on R&D and innovation, and educational statistics. Bibliographic information is held in a private database, which is not unusual. Thomson Reuters, after all, is also private. We have good data on higher education, although this is inadequately exploited, and we are busy building a new database called the Research Information Management System, which will hopefully lead to better research management, both in our research councils and in the universities.

What about data held by regulators? We have a clinical trials register. We have gene banks. We have data on plant breeders' rights, biodiversity, and indigenous knowledge compliance, and we have an ethical clearance system built into the funding awards process. We also have the Promotion of Access to Information Act, which allows access to this information. The Patent Amendment Act has introduced some potential dysfunctions into the system, in that it might well constrain people from taking out patents in South Africa as opposed to exporting their knowledge abroad, which is the exact opposite of what the drafters of the act intended.

We are also data rich at the sector level. We have the South African Earth Observation Network. We have radio astronomy data; seismic data; oceanographic, geological, and meteorological data; social science data; and biodiversity data, including about aquatic diversity. We have an extremely able statistics service.

The question is, who gets access to this and how can it actually be used? Regrettably, there is fragmentation by default. Although Statistics South Africa has a mandate to coordinate national surveys, that is as far as it goes. The national Department of Science and Technology (DST) has a mandate to coordinate science budgets, but not to coordinate information. Indeed, it has experienced extreme difficulty even in the attempt to coordinate science and technology budgets across government. Silos rule in the government domain. Originally the DST mandate was to coordinate R&D budgets. It turned out to be impossible, so it was widened to cover science and technology. That has been even more difficult.

We have a National Advisory Council on Innovation, but it lacks the authority to carry out its mandate. It is currently undergoing its third review. We have the problem of management information systems being designed for one task, but being forced to address other tasks; there are resource limitations of inadequate metadata and training.

So, what are the prospects for coordination? I would like to be optimistic and say they are reasonable and perhaps improving. There is a commitment to monitoring and evaluation at the highest level. We now have in the presidency a minister for monitoring and evaluation. That minister has required each of the ministers of state to enter into an indicator-based performance agreement. This, unfortunately, might bring about perverse behavior. If you are asking me to account for myself, it is in my interest to set the bar as low as possible, because then I will be sure to succeed, and I will get a pat on the back.

We also have a commitment to big science, and that necessitates a lot of work on data coordination. South Africa has given strong support to the African Union's work on science and technology policy, in the form of promoting high-speed, wide-area networking and data sharing, and supporting the development of suitable infrastructure. We also have a commitment to the Organization for Economic Co-operation and Development guidelines for access to publicly funded research. A review of the science system is currently under way, which provides an opportunity to accelerate progress.

I must also draw your attention to the United Nations Educational, Scientific and Cultural Organization (UNESCO) study on the social sciences, *Knowledge Divides,* which was published in 2010. It looks quite critically at the issue of social data.

To conclude, I would like to focus on the North-South relationship. I parody this relationship as one of academic hunters exploiting data gatherers. With due respect to my hosts: "We from the North are the hunters, and you in the South are the gatherers. Collect the data, and we will process it for you. We will keep the datasets, because we cannot trust you to manage or to exploit them properly."

These kinds of interactions lead to mechanisms that restrict inquiry. There are many African countries that have been exploited because they have very interesting archaeological or anthropological resources, and people from other countries wish to come in and get their Ph.D.s, carry out interesting research, and get the credit for that. As a result, hardly surprisingly, many countries have closed their doors. In addition, these unbalanced relationships are used to restrict other kinds of research, particularly social science research that might turn up unwelcome findings.

Finally, there is a large amount of information that has not been digitized, and therefore is not easily accessible. The promotion of legislation for data curation and archiving, and for open-access publication, would be desirable as well.

7. Access to Research Data and Scientific Information Generated with Public Funding in Chile

Patricia Muñoz Palma
National Commission for Scientific and Technological Research, Chile

INTRODUCTION

The progressive and significant increase of public funding for the development of scientific research in Chile, while advancing new technologies and international institutions, impose a new challenge: to insure access to and promote the reuse and preservation over time of scientific and technological data generated using public funds. I will describe some policy initiatives addressing to the access to research data and scientific information that are currently under study in Chile.

The situation in Chile can be summarized as follows: Over 80 percent of research activity is supported by public funding, and these funds are oriented mainly to researchers, universities, and research institutions. The results of the research are published in scientific journals and proceedings, and are available in both electronic and print format. Most research data are not available for access or consultation, and commonly are managed by the researchers themselves.

In order to design and implement a policy initiative addressing the access of data, besides the practices of researchers and institutions, other aspects must be considered, such as legal frameworks, agreements between beneficiaries and funding agencies, intellectual property, copyright, and transparency law, and international recommendations of institutions such as the OECD, ICSU, and CODATA.

In this context, CONICYT (National Commission of Scientific and Technological Investigation) takes into account the country's needs to have an adequate infrastructure and to generate policies that facilitate access to research data and scientific information. For this reason, we have developed the following three main initiatives, as described below.

THE FIRST INITIATIVE: A Program for Management of Research Data and Scientific Information

The principal objectives of the proposed program are to manage, strengthen, and guarantee access to data and scientific information collected and produced in Chile from public funding, in order to:

- Facilitate and promote access to, and exchange of, data and scientific information generated in the country;
- Preserve the scientific patrimony (i.e. specific elements of agreed national importance);
- Promote utilization of international standards for management of data and scientific and technological information in public and private institutions;
- Create information products and services as added-value tools for scientific, productive, and economic development of the country;
- Systematize the process of management of data and scientific information with the goal to obtain data, which will allow the creation and improvement of national indicators regarding investment, transference, and impact of public investment in the national system of science and technology.

The program has considered four strategic items:

1. Institutional framework: Respond to the needs of the country with a broader vision for development of science and technology and innovation.

2. Specific studies: Generate basic inputs for decision making and generation of policies.
3. National and international linkages: Create incentives for development of national and international networks.
4. Human capital: Develop instruments allowing for specialization of professionals in these areas.

The most important product of this program will be the design of a National Platform for Access to Science, Technology, and Innovation Knowledge. This platform will allow consulting and access to research data and information, and will enable their storage and preservation. This platform will involve all the relevant institutions, such as universities and technology research centers, participating in the process of production of technological and scientific knowledge, including the collection of data, processing, production of specific reports.

THE SECOND INITIATIVE: Study State of the Art of Access and Management of Research Data and Scientific Information Generated with Public Funding

This initiative was the first step for the implementation of the above-mentioned program. Its goal was to gain basic understand about international and national practices on access to data and information.

The specific objectives of this study were to:

- Identify the international state of the art on these topics;
- Describe the practices and policies for access to research data and scientific information in Chile; and
- Generate recommendations related to management of research data and scientific information for the Chilean institutions, considering the international context.

This study covered several deficiencies of the Chilean practices, such as the ones explained in more detail below:

Institutional framework:
- Lack of institutional awareness at all institutional levels, about the relevance of keeping adequate data management practices.
- Responsibility for the management of data and scientific information is spread throughout organizations, lacking one specific organizational unit in charge of it.
- Data management within the organizations is financed through individual projects, rather than considered an activity that requires continuity, therefore a permanent source of funding.
- Patrimony is not seen as an institutional and national asset, but as researchers' personal resources.
- Lack of standardization of data and the formats of scientific information.

Human capital:
- Lack of critical mass of human resources specialized in data management.
- Scant supply of formal and specialized academic programs focused on data management.
- Lack of incentives to create upward career paths in the field.
- No leadership or national references in the field.
- Unstable and short-term work positions.

Patrimony:
- Lack of or weak valuation of patrimony.

- Patrimony scattered throughout organizations.
- Patrimony considered personal, rather than institutional or national.
- Preservation and formatting according to individual criteria.
- Lack of initiatives for sharing or exchanging.
- No culture of giving access to patrimony.

Policies:
- Lack of institutional policies addressing access to research data and scientific information.

THE THIRD INITIATIVE: Development of a Policy on Management of Research Data and Scientific Information

The goal of this policy for the main Chilean public funding institution (CONICYT) will be to implement a series of standards for capture, registration, and management of data and information systems to be by all beneficiaries of public funding.

The principal objectives of this activity are to:
- Optimize and rationalize use of public resources involved in generating and managing scientific knowledge;
- Increase access to research data and scientific information; and
- Adopt and attain international standards, including OECD recommendations.

In order to implement this new policy for CONICYT, the next steps will be to:
- Generate awareness in the scientific community, and among academics, public employees, and public agencies about the relevance of having access to research data and scientific information; and
- Develop a network of institutions and individual professionals interested and willing to collaborate to the implementation of this policy.

CONCLUSIONS

These initiatives represent a significant challenge for CONICYT. As the main Chilean funding agency for scientific and technology research, it is aware of the value that the access to research data and scientific information adds to the process of knowledge production. The implementation phase that comes next will encounter resistance at different levels, from legal barriers to deeply established notions of ownership of the data produced by researchers. We trust CONICYT will be able to gather the support of the scientific community and political leaders, and be able to align them around the goal of boosting the quality and quantity of the national scientific and technological production.

8. The Management of Health and Biomedical Data in Tanzania: The Need for a National Scientific Data Policy

Leonard E. G. Mboera
National Institute for Medical Research, Tanzania

Sound statistics are a key component of evidence. This is the main reason that we see that scientific research is increasingly dependent on successful outcomes of access to data. In particular, the research findings that have been conducted in one place can be well utilized or relevant somewhere else. Many developed countries have created some initiatives whereby they have established national policies and programs for data management and access, but the situation is different for developing countries. This includes Tanzania. Most of those countries either have a weak or nonexistent policy for management of data and access.

However, there have been some initiatives taking place in the African region that are worth mentioning. For example, the Algiers Declaration (2008) aims at improving

- The availability of relevant and timely health information, and access to global health information;
- The management of health information through better analysis and interpretation of data;
- The availability of relevant, ethical, and timely research evidence;
- The use of evidence by policy makers and decision makers;
- The dissemination and sharing of information, evidence, and knowledge; and
- The use of information and communication technologies.

As for the current situation in Tanzania, we have developed guidelines on data transfer, but there is no policy on research data sharing yet. Difficulties also exist in accessing and sharing the scientific data that has been collected by different researchers. Specific barriers to the access and sharing of scientific data collected by researchers using either public or donor funding include scientific and technical, institutional and management, economic and financial, legal and policy, and normative and sociocultural aspects.

The National Institute for Medical Research (NIMR) plays a great role in health research and data sharing in the country. For example, the Parliamentary Act No. 23 in 1979 gives the institution a mandate to work on some key issues: (1) to establish a system to register the findings of medical research carried out within Tanzania, and promote the practical application of those findings for the purpose of improving or advancing the health and general welfare of the people of Tanzania; and (2) to establish and operate systems of documentation and dissemination of information on any aspect of the medical research carried out by or on behalf of the institute. This is the basic mandate that the NIMR has had since its establishment more than 30 years ago.

Regarding management of health research in the country, the situation is not very good. Researchers in Tanzania, like in other developing countries, lack the norms and traditions of open data sharing for collaborative research. If you go to individual institutions within the country, you will find that many institutions are treating their data either as a secret or as commercial commodities. They are not really open to say, for example, "Here's the data. Let's do something with it. Let's make sure that we inform the world about the research we have been doing." Also, although the government does not actively protect such data, it lacks policies that provide guidance or identify responsibilities for the researchers in making research data available for others to use. Moreover, Tanzania does not have a central data center or digital repository in place where researchers can submit their data for use by others. This is a key challenge. We are saying that researchers are not willing to submit their data, but if they are, where are they going to put it to share with other institutions or with other people?

Next, I would like to talk about the issue of research data transfer in the country. What are the guiding principles for data-transfer agreements in Tanzania? Tanzania introduced a procedure for data transfer between Tanzanian and foreign institutions in 2010. The Data Transfer Agreement (DTA) is used as the only legal document by institutions in Tanzania to regulate the uses of data they provide to specific research projects inside and outside the country. It also provides opportunity for institutions to claim co-ownership of the improvements made from data the recipient has acquired. Upon approval of the DTA, the agreement becomes valid and the recipient is granted unique access for a period of time, depending on the duration of the project, after which the data are placed in the public domain.

Before concluding, I want to inform you about publication procedures in the country. No research work can be published without getting permission from the National Institute for Medical Research (NIMR). The procedure is simple. When you want to do research in the country, you need to get ethical clearance from the NIMR.

Finally, I want also to tell you about some new initiatives within the Institute for Medical Research. The NIMR is in the process of establishing a Web-based National Health Research Data Repository. The primary purpose of this project is to provide a storage database for all health research work that has been conducted in the country. We want to put together all research work conducted in different places in the country in a central database, whereby it can be accessed.

This has not been easy, and we still have a number of challenges. The financial challenge is the big one. Also, issues of human resources are a challenge, especially regarding technical staff. Furthermore, we have challenges related to infrastructure. We did not give up, though. The NIMR has different centers and stations around the country. Each station or center has a purpose for conducting particular research. Some centers focus on malaria; others focus on tuberculosis or neglected tropical diseases. The data repository for the NIMR will collect information from these centers and stations. Once we are done with them, we can move on to other institutions.

To conclude, with the increased need for data- and information-sharing globally, it is time now for Tanzania to develop and implement a national research data policy. This policy, among other things, will oversee the generation, storage, data transfer, and use of research among local, and between local and international, users of different research findings. Also, we need to have a formal mechanism established to govern the accessibility and use of created research data.

9. The Data-Sharing Policy of the World Meteorological Organization: The Case for International Sharing of Scientific Data

Jack Hayes
U.S. Permanent Representative to the World Meterological Organization,
NOAA Assistant Administrator for Weather Services, and
Director, National Weather Service

The mission of the National Oceanic and Atmospheric Administration (NOAA) National Weather Service is to use the information that science has equipped us with to provide forecasts and warnings to protect life and property and enhance national economies. This will be the focus of my talk.

From the U.S. perspective, we have long believed that free data sharing is a benefit to the United States, and from a perspective of the use of information created using the U.S. tax dollar, it must benefit society and must be made available to people. My focus will be on how it benefits the United States today. I will conclude with some additional thoughts about where free and open access can benefit capacity building in developing countries, and then, as a weather service operator, how I see that benefiting society.

If you are not familiar with the World Meteorological Organization (WMO), it is a specialized agency of the United Nations that was created after World War II, in 1951, although it existed in various forms since the 1870s. The WMO has 190 members. Its mission is to focus international partnerships on the collection, production, and exchange of weather, water, and climate information in order to protect life and property, enhance national economies, and preserve environmental quality. There are two important WMO resolutions related to data sharing:

Resolution 40: "WMO commits itself to broadening and enhancing the free and unrestricted international exchange of meteorological and related data and products." This resolution was passed by the WMO Congress in 1995 to provide free and unrestricted international exchange of meteorological and related data and products across the world. In 1999, Resolution 25 established the same basic policy across the 190 members for hydrological data and products.

Resolution 25: This resolution calls for "committing to broadening and enhancing, whenever possible, the free and unrestricted international exchange of hydrological data and products, in consonance with the requirements for WMO's scientific and technical programs."

The way this policy is implemented across the globe is through giving the national owner of data and products the ability to discriminate between the data that are essential to the core mission—to protect life and property—and the data that are called supplemental. Generally the protect-life-and-property data and products are free and open, and supplemental data have variable degrees of openness, depending on the country.

U.S. data policy has its foundations in the Office of Management and Budget's Circular A-130: "…government information is a valuable national resource, and … the economic benefits to society are maximized when government information is available in a timely and equitable manner to all." What we have found in the United States is that free and unrestricted data sharing maximizes the value for the information we collect and create within the National Weather Service. We also find that the reverse is true. Where we have barriers, we underutilize the data that we collect and produce. Fundamentally, the more people use our information, the more value it has.

To fulfill our mission to protect life and property and to enhance the economy, we collect and produce information that can be used to do that, but this data can be used for other purposes as well. There is a commercial weather sector that uses our data, adds value to it, and provides products and services for very specific needs. This, we find, has aided economic growth within the United States.

Let me give some specific examples. When an F5 tornado hit the community of Greensburg, Kansas, in 2007, the National Weather Service detected it. The tornado occurred at night. It formed in northern Oklahoma and moved across the border into Kansas, where it leveled Greensburg. While the community expected hundreds of people to lose their lives, the number was about 10. This was possible because the National Weather Service detected the tornado and, in partnership with the media, alerted people via television, radio, and NOAA weather radio, so that they had plenty of advance warning. It was a partnership that we had with the emergency management community that worked with the hospitals, so that we had moved ambulances outside of the town that was going to be struck to a position where they could be of more use.

What did the commercial weather sector do in this incident? At that time, there was a train headed right into the path of that tornado outbreak. There was a commercial weather company that used the information we collected to relay a warning to the dispatcher in Omaha, Nebraska. They put a hold on that train several miles east of the path that we projected for the tornado outbreak. That train stopped and the company saved the lives of that train crew and the train itself.

To emphasize the value of data sharing in this incident, I would like to compare it with a very similar tornado outbreak in the early 1950s in Udall, Kansas. In that incident, a very similar tornado moved out of Oklahoma into Udall at night and destroyed the city. As a result, 100 people lost their lives. I would say that the free and open data exchange, and the partnership created between the private and public sectors in the United States, has allowed us to reduce the loss of life over the past 30 to 40 years.

Here is another example. There are various motor raceways across the United States. In the partnership we have with the private sector—for example, the Pocono Raceway in Pennsylvania—our job is to monitor severe weather and to ensure that the raceway has the warning it needs to get the people attending out of harm's way. The private sector will use information we collect about temperature, humidity, wind—anything that might affect the conduct of the race or the comfort of the people there, add value to it, and give it to the raceway officials, to the racers, and to the people in general. That is a partnership that increases the value of the information.

In one last example, in partnership with the Federal Aviation Administration and the Department of Defense, we have roughly 150 Doppler radars spread across the United States. We use the data from these radars to detect severe weather and to warn others of severe weather. The commercial weather industry provides applications. If you have a BlackBerry or a cell phone that has an imagery capability, you can get access to that imagery.

I will switch to a discussion of Europe now. Different countries in Europe have different models. In general, however, many countries supplement their federal budgets by charging for some data and products that they produce. Protect-life-and-property data and products are generally freely available, and their warnings and observations are too. The warnings, in general, are simple cartoons, not as vivid as what we can produce in the United States. The observations are generally conveyed as text rather than as a visual representation.

Let me talk about capacity building in developing countries and give you an example of where a united effort sponsored by the WMO is certainly going to enhance the capacity within developing countries. In the wake of the December 2004 tsunami in the Indian Ocean, nations of the world got together, through the WMO and the International Oceanographic Commission, and created a tsunami warning system that is used internationally. It is that free and open exchange of information that aided and helped us move that along. One effort that I have been personally involved in is in southern Africa, called the Severe Weather Forecast Demonstration Project. I have long objected to the use of the word "demonstration," because basically it means that we take model information based on geophysical equations and apply it to predicting and warning citizens. I do not think we need to demonstrate that; we have seen that for 30 or 40 years in the United States. It is a matter of growing actual predictive capabilities in developing countries.

What we have done is allied with other similarly inclined developed countries and created in South Africa the ability to bring in global model data. The government of South Africa runs a regional very-high-resolution model, combines it with satellite information, and creates products and data that it distributes to 16 developing countries in the Southern African Development Community. Similar initiatives have been undertaken in East Africa, Northwest Africa, the Southwest Pacific, and Southeast Asia, as well as the Caribbean. These initiatives are regionally driven, in which a subregion or a region makes a statement about its environmental threats, needs, and priorities. Under the WMO auspices, developed countries help to develop capacity so that we can transition to those countries things that we have become used to in the United States.

Looking at a world picture over the past 50 years, increased data sharing has contributed to a major reduction in loss of life. In the 1956-to-1963 time frame, 2.6 million people lost their lives due to environmental disasters related to weather, water, and climate. Fast-forward to 1996 through 2005, when 220,000 lost their lives. This is still too many, but it is an order-of-magnitude reduction in loss of life due to weather, water, and climate.

Looking to the future, initiatives such as the Global Earth Observation System of Systems (GEOSS) is an attempt to create data sharing for future scientific and operational benefit. As we look to the nine societal benefit areas in GEOSS, we are already seeing the scientific interrelatedness. It will only grow. From my perspective, free and open data exchange helps us advance the vision of GEOSS faster, cheaper, and further than we could if we did not have that capability.

To conclude, from a U.S. perspective, free and unrestricted data sharing in our country has created great value that we realize today. We strongly support initiatives to sustain and broaden it, both in the United States and internationally.

10. DISCUSSION OF PART TWO BY THE SYMPSOSIUM PARTICIPANTS

PARTICIPANT: I have a question for Dr. Hayes. When you mentioned the less-than-optimal policies of certain European countries in data sharing, do you see an improvement in recent years in that respect?

DR. HAYES: The answer is yes. In the 1990s, there was very strong opposition to more free and open data exchange policies. However, there is a trend in many of those countries today to adopt our model, if not formally, at least informally.

PARTICIPANT: My question is to Dr. Kahn. Do you have any divisions between locally published articles versus internationally recognized journals—*Nature*, *Science*, and even specialized journals?

DR. KAHN: That is indeed the very problem of utilizing Thomas Reuters or Scopus or any international repository to measure publication output. It is naturally biased in three ways: language, field, and against locally published journals. It is a restricted view, but it has become the gold standard. Garfield gave us the ISI nearly 50 years ago, and it has retained its status up to the present time, but it has to be used with caution, which is why I made the remark I did. If we take any country outside the core of the Organization for Economic Co-operation and Development (OECD), there will be a host of local journals in the local language, particularly in the social sciences, law, economics, and so on, that are not counted. To get a true picture of the contribution to the knowledge pool, you have to extend what is included. We can assume, of course, that these local journals are subject to peer review and have an appropriate frequency of appearance. Plus, in the social sciences, publication channels differ. Books are more common than journal articles.

PARTICIPANT: I have one partial correction for you and then a question. The correction is that the implementation guidelines that you showed for the Global Earth Observation System of Systems (GEOSS) are no longer proposed; they are actually accepted by the GEOSS plenary. That is progress. The group is still working on pushing that further. There are a lot of issues that those guidelines and the original principles do not address.

My question is that there were a lot of very interesting presentations about the responsibilities of the developing countries' governments to coordinate and get their data policies and act together. We have heard less about what the responsibilities of developed countries should be, having had a tradition of being hunters. From your perspective, how should developed countries behave? What practices should they adopt for acting respectfully toward developing countries' efforts?

MR. MAYALA: I think that there is a need for the developed countries to focus on what developing countries are doing. Of course, if there is any support they can provide, then they should provide it. We need to learn from each other. From there, we can move together.

DR. BALSTAD: A few years ago, the International Council for Science sponsored a study of various priority areas in international science. One of them was data and information. The report that came out on data and information was not intended to be a manual or a policy prescription, but something in between that discussed major issues in data and information and some of the responses that have been developed in the developing world. Part of it is that we are all learning in this new area, and we need to work together. Some countries have gone farther down this road than others. It is vitally important that they share information about how to do it.

DR. KAHN: Remember the medical oath of Hippocrates: First, do no harm. That is what doctors agree to do when they are sworn into the profession. In asking developing countries to make data available, we should expect that there would be reciprocity from the North. The reciprocity should include an understanding of the defensiveness that might be encountered from the South, which is a residue of many years of a lopsided set of relationships. A book like John Le Carré's *The Constant Gardener* does not come out of nowhere. There is lopsidedness and often a legitimate concern that we in the South are, as one cabinet minister put it, a giant zoo for the benefit of others. The important thing is that the barriers need to be negotiated away, lest the general scientific enterprise suffers.

PARTICIPANT: I would like to share some of our experiences and the challenges that we face with this. Our project does computer modeling of infectious disease transmission, and for that, we typically use high-resolution surveillance data from governments. For this, we tried to form partnerships with countries in Southeast Asia (Laos, Cambodia, Vietnam, and Thailand), and in Africa (Niger and Nigeria). Our approach has been to make agreements with ministries of health to use data collaboratively and then build capacity inside the ministries to manage data, where needed, digitize paper records, and then put them up in repositories that can be used, and also train people in modeling. In some countries, like Thailand, that has worked very well. In other countries, we can make agreements with ministries to share and coauthor papers and provide technical reports, especially in very-low-income countries, but our experience has also been that the time that people have in ministries of health or in governmental agencies is incredibly limited to interact with us and to be trained. They have very limited capacity to train people and then use the data management system at all. So it has been a struggle for us to identify people and push governments to see if there are people available to be trained. There is also the issue of lack of interest in the data itself. We found that a lot of governments do not really seem to care too much or do not see the value yet. In summary, our challenge is the sustainability of training people and making sure that after we leave, the country retains the databases and data systems and modeling capacity, and it has also retained the interest of the people locally.

PARTICIPANT: My question is for Ms. Muñoz. You gave us a very comprehensive description of the initiatives of the National Commission of Scientific and Technological Investigation (CONICYT) with regard to promoting open data-sharing policies, but I wonder whether you could expand on the efforts of CONICYT to promote a culture of data sharing with the international community for Chilean scientists.

MS. MUÑOZ: I think that we can work with the whole community to promote open access to data and scientific information, but at this moment, we think that the most important initiative will be to develop and implement the policy for management of publicly funded research data and scientific information, because this is the missing gap.

PARTICIPANT: My question is about how you see the new paradigm coming out from the cloud-computing utilities that, in a way, the scientific community is going to adopt. We can see that this phenomenon is growing very fast in the United States. The National Oceanic and Atmospheric Administration (NOAA), for example, is leading, with other agencies, this kind of phenomenon. I am very interested in your perspective. The second question concerns how you see this new utility from the technological point of view, as a way to boost collaboration among developing countries. I am representing Italy, and am very interested to see how we can boost this transatlantic dialogue between the United States and Europe in order to support developing countries.

DR. HAYES: From a NOAA perspective or a National Weather Service perspective, I put on rose-colored glasses and say, "How could we benefit society?" Then I take the glasses off and say, "What constraints do I have to live with?" One of my challenges here is to increase the value of weather information to the United States. I think that creating a concept like an open National Weather Service—which would include the public sector, the private sector, and the academic sector—would ensure that. This allows research managers to talk with operations managers and encourages data-sharing partnerships among a variety of missions.

Given the importance of water to our future, we have launched an integrated water resources science and services within NOAA. There are 21 federal agencies in the United States government that produce water information. I would think that we could partner across the U.S. federal government, share data and information, and increase its value and partnership to the United States. I would broaden that to the private sector as well.

Another example is one that we have just started. It has to do with space weather. We are going to have a solar maximum in 2013. The infrastructure that we have created is more technologically advanced, but it is also more vulnerable to things that happen on the sun than it was even 20 years ago. I envision a future where we could create an international partnership around the globe through an organization like the World Meteorological Organization (WMO) to bring together academic, operational, and private-sector resources to benefit society.

PARTICIPANT: I would like to address a question to Dr. Hayes about what might be related issues and possibly also get some response or input from our colleagues from Africa. Is there any information available about trends in the participation of developing countries in the WMO GTS (Global Telecommunication System) weather system? Do you see more data coming into the WMO from developing countries, particularly Africa? Also, has the WMO become more active in recent years in providing services or products, above and beyond the basic weather forecasting, that would be of special benefit for developing countries? Again, I ask particularly with respect to Africa.

DR. HAYES: I will try to address it, and if I miss something, you can come back to me. First, observations are not inexpensive. You can imagine a developing country has similar challenges. We also have commercial interests that are threatening; for example, 4G mobile communication standards are invading the space that used to be reserved for meteorological sensing and data transmission. Second, going beyond weather, there is a concept in WMO called the Regional Specialized Meteorological Center. It really, I think, stemmed out of Chernobyl, where some countries had a nuclear-dispersion capacity and could share that with regions and make those products available. I certainly see that capability. As it applies to Africa, I think the Severe Weather Demonstration Program that WMO is using is trying to anchor within a sub-region of Africa in more than just the computing and distribution infrastructure, but in the ownership as well.

PARTICIPANT: This is a question for panelists Muñoz, Mayala, and Kahn. In developing your respective national science data policies, did you find the greatest resistance coming from political forces, economic or business forces, the scientific community, or just plain inertia? In other words, we have not needed it before; why do we need it now?

DR. KAHN: Previously, I worked for our Human Sciences Research Council at a time when we were busy debating the meaning of the OECD 2007 guidelines, which are great and noble and agreed upon by all. The Human Sciences Research Council, where I worked, was a research performer, unlike the Economic and Social Research Council of the United Kingdom, which is a grant maker. The parody at the time was that our universities wanted only two simple things from us: all of our research funds and all of our data. The parody was, you show us yours and we will show you ours. It is a very critical issue. We are moving toward an open-access data policy. Certainly the national Department of Science and Technology believes that this is a good thing, but at the same time you get what I will call a "techno-nationalism" that creeps in. It is even present in what I heard this morning from Tanzania. We have been cheated in the past, and therefore we have to be on guard to ensure it does not happen in the future. So you get these countervailing tendencies. On the one hand, there is something we might even call data sovereignty. On the other hand, there is data openness. We all agree that data openness is the way to go, but in the evening, when we have a drink with our party colleagues, we say, "No, no, no. The nasty foreigners are going to take it all away from us, and we have to be on our guard all the time." It comes from two sides.

You asked where most of the resistance comes from. I think there is a political dimension. The business sector is another case, because in South Africa the greatest volume of research and development is done by the business community. In that sense, we are more like an OECD core member state than an emerging economy. However, the business community does not publish very much in the scientific literature, so that does not really come up. The complication arises when you have a research project, and this takes us to the Bayh-Dole Act in your country—a research project that is jointly funded by business and public funds. How do you now decide which piece of data goes into the public domain and which remains behind the company walls? There is resistance. It is political. It is nationalistic, techno-nationalism, as I said. There are also proprietary restrictions.

MS. MUÑOZ: I think that the big problem for implementing this in Chile is our scientific community, because of the cultural aspect. In this context, the researchers do not have knowledge about what is important in these matters in the development of the country, and its public value.

MR. MAYALA: I want to give a little bit of experience from my country. First of all, I think it is known that you cannot put science and politics in the same pot. It is very difficult to act together. In my country, we want to come up with something that people will see. There are those groups who wait and see and then they react. We are in that kind of situation. I mentioned that we are trying to build a data repository. We hope that when it is out, we can get an appropriate reaction from people.

PARTICIPANT: I have a question for Dr. Kahn. You mentioned that South Africa is now a member of the BRICS Group. I presume that means more cooperation and partnerships with the large developing countries that are members of that network. You also mentioned that there is the issue of South Africa in Africa. You did not talk about that as much. Could you describe some of the partnerships and networking that are going on in South Africa with other African countries?

DR. KAHN: First, on an economic standpoint, South Africa accounts for something like 30 percent of the continent's gross domestic product and around 70 percent of its scientific production. Those two walk together. South Africa, for the last 10 years, has been paying the bill to host the New Partnership for Africa's Development and now the African Union's Science and Technology Secretariat, which is hosted in our Council of Science and Industrial Research in Pretoria. In addition, I mentioned the role that we are playing in supporting higher-education development. I mentioned a figure of 50,000 students from across Africa studying in South Africa, out of a student population of around a half a million. It is close to 10 percent. Of these 50,000, fully 30,000 come from an economic grouping known as the Southern African Development Community of 15 southern African countries, including Madagascar. Those students are regarded as home students. They pay exactly the same fees as I pay for my own children. So, South Africa is donating 1 billion Rand (the equivalent to 122.1 million U.S. dollars) a year to support students from the Southern African Development Community. This is very important.

Because of the economic dominance of the country, there are many international agreements to which we are a party and where we play a central, coordinating role, ranging from air traffic navigation to weather and climate mapping and the like.

Southern Africa's achievements lie in environmental modeling and monitoring cross-borders, as well as in the harmonization of transport links. All of these involve considerable research and data sharing. So we in South Africa are playing a role both in capacity development and in data sharing in numerous fields, across southern Africa in particular, but further afield as well. A lot of this, of course, is driven by self-interest, and our business activities to the north are well supported by this research investment over many years.

PARTICIPANT: Some international organizations still have contributing members who are not following those data policies. How can we provide incentives rather than simply just examples as we move forward? Are there lessons to be learned from other communities that have had success?

DR. HAYES: I will speak from a meteorological perspective. One of the challenges in our community is that ownership of the data is country dependent. The philosophy we use in the United States is not the same philosophy they use in Russia. It is more than data. It is how the government is organized and how they fund the services. While we may not feel a threat in the United States, the Russian Hydrometeorological Service does. You have to almost deal with those one at a time, country by country, and be willing to sit down and try to find a way to reduce the perception of threat.

**PART THREE:
COMPELLING BENEFITS**

11. Developing the Rice Genome in China

Huanming Yang
BGI (Beijing Genomics Institute), China

In this presentation, I will share with you information about the development of the rice genome as well as other genomes in China, through international collaboration and international sharing of scientific data.

The field of genomics was cultivated by the Human Genome Project (HGP). It was important to China that its scientists made a contribution, even a small one, to the HGP.

China was a latecomer. One of my colleagues wrote: "The work was already underway in other countries, and China was way behind the curve. But Huanming convinced me that China's involvement would represent a major advance for his country and for the Beijing Genome Center." Our idea was that with just a simple contribution from us, together with all other countries, we could benefit from this project as could all the people in the world. It was so expensive that China alone could not afford it. That was the reason for us to support international scientific collaboration.

As a member of the International Bioethics Committee of UNESCO, I strongly proposed that the most important and urgent issue in bioethics at present is the immediate release and free sharing of the human genome data. In a period of 5 months, I submitted four proposals for free sharing of human genome data. Then on May 9, 2000, the director-general of UNESCO issued a statement on the free availability of human genome data. This led to the Group of 8 (G8) communiqué of the Okinawa meeting in July 2000, as well as the United Nations Millennium Declaration on September 19, 2000, to ensure free access to information on the human genome sequence.

As a unique member state from the developing countries in the HGP consortium, China's contribution is not only a technical accomplishment, but also a recognized effort in the free sharing of human genome data. As Michael Morgan, who was responsible for the Wellcome Trust projects on genomics, said, "China's unswerving support of open data release was an important factor in ensuring that the human genome sequence is the property of the whole world." John Sulston, the leader of the HGP of the United Kingdom, said, "I especially salute the Chinese colleagues, who have affirmed the Human Genome Project's common ownership by all mankind."

The Human Genome Project has established a brilliant example for international collaboration and data sharing. With no participation in the HGP or no international data sharing, genomics in China today would not be so advanced.

With regard to the rice genome data, we freely shared and published the first draft sequence of the rice genome in April 2002. It was a big event in the history of natural science in China. That is the reason that the editorial department of *Science* magazine went to Beijing for the news release: "*Science* magazine honoring China's sequencing of the rice genome." I would like to quote from *Science*: "This team deserves enormous credit for their outstanding world-class accomplishment in a remarkably short time."

It is true that the whole job was done in 74 days. Then we released all the data. Our database on the rice genome has been one of the most popular databases in global genomics. If regarding citations as one indicator of the impact of that paper, we are proud to see that our paper has been cited about 2,000 times up to now, and the citations are still increasing. The number of specific publications in one field could be another indication of the impact of the work in that field. Since the release and the sharing of the rice genome data, the research on rice has dramatically increased and has outnumbered that of wheat, though they were more or less the same at the starting line.

For our students, it would be difficult to imagine how research could be done on rice without referring to its genome sequences. This is also the case for all the researchers in the field of genomics. I will give you two examples showing how significant the impact of the free sharing of the rice genome sequence can be.

We are now collaborating with various institutes and universities within and outside China to sequence 10,000 rice accessions or strains. We believe that is an important stimulus for rice genome research and breeding. We also have analyzed the genomes of other important crops, such as maize and potatoes. A paper about this research will be published in a prestigious journal very soon. We also have analyzed the cucumber, and sequenced the genes related to sex expression, disease resistance, and so on. As soybean curd is an important food for the Chinese, we have also completed the genome analysis on soybeans.

We have analyzed the genome for chickens. One of the most important scientific discoveries is that the genome diversity of chickens originated before their domestication.

Together with other institutions, we have analyzed the genome for silkworms. The most thorough genome sequence data so far for the silkworm was released by BGI and other Chinese researchers many years ago. Now the job is continued by analyzing about 50 genomes, with the aim to reveal the domestication events and genes for this important insect.

We have analyzed the panda genome. Technically it is the first successful *de novo* assembly of a mammal, without the help from a genetic or physical map of this animal. We have also analyzed, in collaboration with our colleagues in the United States, the genome of two ant species to reveal how they organize their social activities.

We are also concerned about global climate change, so we have finished the sequencing part of some animals living in extreme environments, like the polar bear, penguin, the Tibetan antelope, and the camel.

We have initiated, together with our colleagues all around the world, the International 10K Genomes Project. About 25 percent of all the vertebrates are listed in this project. We have already begun studying the first batch of more than 100 species.

Now I would like to turn back to the human genome research. As I have already indicated, the Human Genome Project has shaped the field of human genomics, which is characterized by international collaboration and international data sharing. Our contribution to the HGP has been small, probably 1 percent, but we made about a 10 percent contribution with the International HapMap Project. We completed the first Asian human genome by means of the new-generation sequencers and published the results in *Nature*. The publication of the Asian genome revealed that the methods or sequencing technology available now are extremely powerful, as stated in a review paper in *Science*.

Perhaps you have seen that my institute is committed to sequencing more than 10,000 individuals before the end of this year. We are also an essential part of the Human Variome Project, as well as the International Cancer Genome Project. For the International Cancer Genome Project, we are committed to gastric cancer now, and we will expand our contribution to other cancers related to digestive systems. We have initiated the 1,000 Monogenic Diseases Project, based on our own experiences working on the Mendelian disorders. One of the most important discoveries in 2009 was the "human pan-genome". We have identified that 0.6 ~ 1.5 percent of human genome sequences actually are population-specific.

We also analyzed people living in two different environments to reveal how their genomes have adapted to the environment. The population study of Tibetans and Han Chinese uncovered that regulation of the hypoxia response is central to high-altitude adaptation. That paper was published in *Science*.

We captured more than 80 percent of the human genome sequence from the hairs of a sample which was at least 4,000 years old. Then, we published the first catalogue of the human gut metagenomes, in which we identified at least 2,000 species of bacteria. Finally, we also have initiated a 10,000 Microbial Genome Project, as well as the Earth Microbiome Project, also through international collaboration.

As scientists, we acknowledge that we have a responsibility to the world. In 1994, I told my colleagues what we should bring back is not only the technology, but also the ethical principles. As co-chairman of the European Actions for Global Life Sciences, we developed this slogan: "To raise the banner of science and humanity." I am president of BGI (formerly known as Beijing Genomics Institute), and the institute's slogan is: "To raise the banner of innovation and ethics."

BGI's mission is to share what we have with others to promote global genomics, to work together with those who are unable to access what they need, and to work together with others to do something bigger, faster, cheaper, and better. It is, after all, so important to build the capacity in developing countries.

I think that international sharing of scientific data is not only an issue of science, nor merely an issue of economics, but also an issue of humanity and global harmony. In genomics, we Chinese already have benefited so much from it. We are told by our ancestors that nobody can be a hero without three partners. Genomics cannot be done alone.

BGI's strategy for sustainable development requires international collaboration and data sharing. None of our achievements would have been gained without international collaboration. We are confident of our passionate young staff, but of course, enthusiasm is not enough for science. Innovation is the key for scientific development. Bioinformatics software is our major innovation. That is how we have made rice genome assembly possible, as well as most, if not all, of the software that we are using for next-generation sequencers.

Nonetheless, humbleness is rooted in our Chinese culture. We began with nothing, though now my institute is poised to become the biggest DNA sequencing lab in the world. When we are called the sequencing factory for the world, we are happy to have this nickname. According to *Science*, "BGI-Shenzhen enhances its reputation as the world's largest sequencing center, deciphering an ant, the Asian human genome, the human methylome, and a gene catalog of the human gut microbes." At the same time, we appreciate again all those who have helped us. Just like the Chinese proverb says: "When you drink sweet water from the well, do not forget who helped dig it."

12. Data Sharing in Astronomy

Željko Ivezić
University of Washington, United States

I am going to summarize a few experiences we collected in astronomy with data sharing. I will first list a few questions that we deal with in astronomy just to set the stage. Then I will review what we learned from the Sloan Digital Sky Survey (SDSS) and similar surveys in the context of data sharing. My main point is that we have to submit ourselves to cost-benefit analysis and see whether or not it pays back.

What we are trying to do in astronomy is fairly well summarized in these three big questions:
1. How and when did the universe begin?
2. How did the structure (planets, stars, galaxies) in the universe form and evolve?
3. Is our planetary system unique (or, is there life anywhere else)?

You can rephrase this by asking if the physics we learned on earth is applicable to the rest of the universe, and if we can use the observations of the universe and of the heavens to learn more physics. These are, so to say, business questions, but we should never forget that, going from families visiting the Smithsonian Institution in Washington, D.C., to a small village in some corner of the world where grandparents tell their children the stories about stars, we are all fascinated by our place in the cosmos.

Over the last decade or so, there has been an explosion in new tools and methods. Due to the fast development of computer and information technologies, we now have new tools, methods, and cutting edge sky surveys that allow us to observe the whole sky in great detail. There are three frontiers in optical astronomy.

The first is to build ever-larger telescopes. These are, for example, the twin Keck telescopes in Hawaii. We build large telescopes not to get more detail, but to see fainter objects. The second frontier is to launch our telescopes in space, above the Earth's atmosphere, which blurs the images of ground-based telescopes and absorbs all the light except visible and radio. The beautiful images from the Hubble Space Telescope are so detailed not because Hubble is exceptionally large, but because the images are not blurred by the atmosphere.

Both the Keck telescopes and Hubble more or less study one object at a time, but they cannot see the whole sky. In fact, all of the area in the sky that Hubble imaged to date is less than one-thousandth of the whole sky. With new technology, we can begin to cover the entire sky and get diverse and precise data on hundreds of millions of objects. That progress brought in the third frontier in astronomy: gathering digital data for these large numbers of sources and then using statistical analysis and data-mining methods to study them.

One of the key developments related to this symposium is that about a decade ago astronomers started sharing all these giant databases freely with the public. What do these data contain? First, they have images. Observers are interested in the position of the object in the sky, its brightness, the size and shape of its image. Objects are observed with different filters to get information about the spectral energy distribution. Images are processed to measure objects, such as stars and galaxies, and to construct catalogues, which are the most useful data products that we put in the public domain—of course, together with images.

Why are these catalogues important? We can use them to discover new objects, classify those new objects, study them statistically, and search for unusual objects. The larger the sample is, the more unusual stars or galaxies will be found. Then we can study cosmology and try to answer these questions:

When did the universe begin? How did it develop?

When astronomy databases are open to the public, more people can participate. One of the foremost examples of this new generation of surveys is the Sloan Digital Sky Survey (SDSS). It was the first time that we could get digital data, a color map of the sky, for a substantial fraction of the sky. There have been a number of projects in astronomy in the last decade that have built digital databases and made them available to the world's public.

SDSS used a camera that has 120 megapixels. It used to be the largest camera in the world. For more than half a decade we collected measurements—positions, colors, shapes, and so on—for about 400 million objects. When the database containing these measurements was released to the public, even before a substantial fraction of science analysis was completed by the project team, that was a revolution in astronomy. This approach is still not accepted in all the fields, but I would argue that there are clear benefits of doing so.

The SDSS public portal provides astronomical data to anyone, anywhere. Two years ago, I was vacationing in Croatia, my country of origin, and I met a friend from Serbia who is an astronomer who had a house on the same island. We got all the data for a paper in four days by accessing this database through our laptops while drinking beer in a beach coffee bar.

The portal also has a special section with many exercises for teachers and supplemental material that shows K–12 teachers how to use these data in the classroom. There are literally tens of thousands of examples where the data were incorporated in school curricula. I would like to remind you that astronomy is very effective in attracting students to science, technology, engineering, and mathematics professions.

As a result of this public data release, again even before a substantial fraction of science was done by the project team itself, several thousand papers were published by scientists not associated with the SDSS—more papers than by the people who were members of the SDSS collaboration. The total data volume that was delivered through this portal was more than 100 times the size of the full SDSS database. There were more than 300 million Web hits over 6 years. The number of unique users was about 1 million. Compare this to about 10,000 professional astronomers. It is a huge impact. As a result of SDSS and previous work, the intellectual father of SDSS, Prof. Jim Gunn from Princeton University, was awarded the National Medal of Science by President Obama.

Over the last few years, even more portals have been developed. One was developed by Microsoft and called WorldWide Telescope. You can use it to get not only SDSS data, but just about every large astronomical dataset. Google Sky did a similar thing. A colleague who developed Google Sky with people from Google told me that as soon as they put it online, there were several million downloads. Google thought that someone was giving them a denial-of-service attack. That shows many people were interested.

What did we learn from this exercise of releasing data even before the project team did its own science? First, it requires higher standards. When you put something out to the public, you have to have documentation. You cannot put out documentation that is not spell-checked, for example. It is a very serious job to put something in the public domain. Then there are costs to keep it public, such as servers, help desks, and so on. These are the two main issues: higher standards and the cost of curation, and because there are costs, we cannot just say, let's do it. We have to subject this idea of publicly releasing data to a cost-benefit analysis.

I will go through a list of the top 10 benefits that we think we extracted from our experience in astronomy. The caveat is, not all of them may apply to your field. Also, they may vary with time. For example, 20 years ago in astronomy, not many astronomers would buy the arguments that I am going to present to you, but today most of them would. Things change with time.

The primary benefit, in the view of many, is that while you are taking data, and you have a finite lifetime for your project, if your dataset is complex, you need to subject it to rigorous analysis to be certain that everything is fine with your data-taking strategy, with your instrument, and so on. Some easy things, of course, you can see immediately. In astronomy, if you get a blank image, then you know that something is wrong. But there are effects at the 1 percent level. With these statistical surveys that have hundreds of millions of objects, we study such percent-level effects. Often they can be hidden systematically in the data that you cannot see with the naked eye; you cannot even see with simple analysis. You need to do complex analysis and cutting-edge research to discover something special with your data. If the data collecting is already finished, then it is too late. That is the main reason why you should do it early, before the end of your project.

Then, especially in astronomy, sometimes you want to take data with different facilities at the same time. In astronomy there are objects in the sky that do not exist forever. Supernova explosions last for a month or so; similarly, there are some asteroids that pass close to Earth, and then we do not see them for decades. There are events in the sky that last a short time and we want to deploy many facilities. To enable this, you have to release your data early.

If you release your data to the whole world, you will have many more users doing your science. With SDSS in particular, outsiders wrote more papers. Many of these papers had great ideas that were not even listed in the project book that was given to funding agencies to justify this investment of close to $200 million. More people mean more ideas.

If you release your data for everyone to work on, you ensure that all the scientific results will be reproducible. They can be verified and you will preserve your data for posterity. That is, again, very important in astronomy. Astronomers strongly believe that code should be released with the data, not just the final product. In particular, in astronomy, because of the atmospheric effects, we see from the ground only optical and radio wavelengths, but we learn a lot about the heavens by observing in X-ray and infrared bandwidths, among others, and using telescopes on satellites. To take this same argument further, cross-disciplinary science is enabled.

There are some fantastic examples from the astronomy–statistics–computer science boundary, where computer scientists use astronomical data to develop new ideas, which are then used once again in astronomy to do better science than astronomers could have done on our own. One can extend this also to other fields, because most disciplines have these giant datasets today. Many of the problems are the same: How do you do data mining in highly multidimensional space? How do you visualize massive datasets? How do you store and query 100 petabytes? These are all common problems. It is not efficient to try to solve them N times when we could solve them together.

Sometimes collaborating and promising to release the data are the only way to secure scarce resources, especially in astronomy where tools are becoming very expensive. Most of the easy things were done in the last century—things you could do in a basement, with one professor and three graduate students. Today's more difficult problems require major resources. Also if you look at the gross domestic product (GDP) of a big country like the United States, in the 1950s, it was 50 percent of the world's economy. Today, it is only one-quarter of the world GDP: therefore, it is now much more profitable for the United States to enter international partnerships than it was 50 years ago. In astronomy, very often when you have a major undertaking, you have to get everyone on board. That is why it is good to share your data.

Going back to the team that produced the data, we are all concerned about our careers, especially when we are young postdoctoral researchers (postdocs) with small children to feed. We want to protect these young people. We want them in successful positions. Indeed, we heard many counterarguments against early release of data from SDSS, claiming that all of us who were postdocs at that time would not get jobs because other people would do the science. What happened in practice was that all of us who were postdocs had know-how about the dataset. We did all the early science, even though the data was public. There was a delay of a year to two for the world community to get on board. Furthermore, our know-how became a marketable commodity, so all the young postdocs who worked on SDSS—literally, all of them, a few dozen—are faculty today.

Then, especially in astronomy, but in other fields as well, education and public outreach are important aspects of our science endeavors. You may have heard about the Galaxy Zoo project, where members of the public visually classify galaxies from the SDSS. The project has already recruited 200,000 volunteers, who went through images of a million galaxies. Evidently, the general public is greatly fascinated by astronomy.

Finally, there are issues of ethics and broader impact. First, sharing is nice, as we all know from our kindergarten days. Furthermore, taxpayers paid for most of our projects. They have the right to see the results at every level of understanding, looking at astronomical images through Google Sky and WorldWide Telescope, all the way to seeing scientific results and translating these to understand better how the universe began, and on the other big questions.

A big aspect of sharing data worldwide is that we are enabling democratization of scientific research. In the context of developing countries, this public release of giant datasets allows small teams to do big science. That colleague of mine and I on the beach in Croatia were a pretty small team, just two of us, and we managed to do cutting-edge science because SDSS data were made public worldwide.

These are the top benefits, but then there are some other issues that we should consider when thinking about releasing data. For example, if you spend some resources to release the data, is anyone going to use them? How many customers do you expect?

Of course, there are different kinds of data. Sometimes they are of great use, like astronomical data. We saw millions of hits to public data releases of people looking at images of the sky. Sometimes data are so specific that even if you release them, not many people would use them—for example, giant datastreams from particle accelerators. That is another extreme.

Then there are security and proprietary issues. For example, a big astronomical project, Pan-STARRS, was done in collaboration with the Air Force. They have to mask some parts of images and not release them to the public. There are also issues of commercial gains and foreign competition.

Jim Gray, who was a Principal Researcher of Microsoft Research Lab, developed parts of those tools that SDSS used to manage and release data. He liked to joke that he loved to work with astronomical data because "they are worthless." What he meant was that they do not have commercial value. With other databases, he had to be very careful about what was released.

If you are doing a major data release that can cost tens of millions of dollars, you have to justify the cost. Once, I listed these top benefits on one page. Being a scientist, I immediately asked myself if there is any predictive value in that list. Can we go through other projects and see if they are consistent with this reasoning? Should they publicly release their data? I will share with you two examples, one from astronomy and one from physics. The first one, an astronomical example, will be a new telescope, a successor to SDSS called the Large Synoptic Survey Telescope (LSST), for which I am the project scientist. Located in Chile, the collaboration includes more than 30 U.S. institutions and about a half dozen European institutions. The shortest way to summarize the difference is that the SDSS gave us the first digital color map of its kind, a snapshot. With LSST, we will observe the sky in a similar fashion as SDSS, but with about 1,000 times greater resolution. Essentially, you can think of it as a digital movie—the greatest movie ever made. If you watched that movie, it would take you a full year without sleep, just staring at the screen. This telescope will have an 8-meter aperture, not a 2.5-meter one, like SDSS. Because its sensitivity will be greater, LSST will detect many more objects. Instead of 400 million objects detected with SDSS, we will collect 20 billion objects with this telescope. It will be the first time that astronomers will have cataloged more objects than living people on Earth. Everyone can get their own galaxy.

When you look at the data volume, SDSS was about 40 terabytes. This is roughly the data volume of the books in the U.S. Library of Congress. Now, when you want to compare LSST, about 100 petabytes, 100,000 terabytes, it is about the same as the volume of all the words ever printed in the world since Gutenberg. Of course, so much data can be problematic. We have 20 billion objects, and we have to track them in real time.

On my list of 10 benefits, indeed, all 10 of them apply to the LSST dataset. It is not a great surprise, because it is so similar to SDSS. But because of this clear win in the cost-benefit analysis, all of the data will be made public to the world. Phenomena that change in the sky will be released within 60 seconds. Then every year there will be a cumulative data release as well (about 10 petabytes of new data every year).

Let me tell you just a few words about another big project in physics, the Laser Interferometer Gravitational Wave Observatory (LIGO). There are two sites in Louisiana and Washington. The project is trying to detect gravitational waves. If it is successful, that is a Nobel Prize-class discovery that would have a great impact on our understanding of the fundamental physics. Should those data be released immediately, or not? It turns out that most of the top 10 benefits apply to LIGO, too, but it is a more difficult question.

To summarize, the issue of public data sharing is a matter of cost-benefit analysis. Often when you do a detailed analysis, you are led to the conclusion that you just have to do it. Given all the parameters of the problem, all the benefits, and all other forces that act upon you, you simply have to share the data.

13. Sharing Engineering Data for Failure Analysis in Airplane Crashes: Creation of a Web-based Knowledge System

Daniel I. Cheney
Federal Aviation Administration, United States

What I would like to address today is an information system that the Federal Aviation Administration has developed regarding transport airplane accidents. It is an initiative to gather information and lessons learned from these accidents, because they definitely have been repeated. The origin of this effort goes back to some problems that we worked on in the late 1990s and the early 2000s. There were several accidents, some of them very large that were heavily covered in the media. When we looked at them carefully, they exposed deficiencies about the way the aircrafts were operated, shortcomings in the maintenance programs, and in the fundamentals of the design of the aircraft. The processes that linked them together were inconsistent. The closer we looked, the starker the inconsistencies were.

Four accidents drew particular attention. These were the TWA Flight 800, Swissair Flight 111, Alaska Airlines Flight 261, and the American Eagle Flight 4184.

From an investigator's standpoint, the TWA Flight 800 accident investigation was a very long, complicated, and technical one. It took years to assemble the wreckage, and much research was done to understand the science behind the cause. At the end, we certainly knew more about fuel system flammability than we did before this accident. Much has been done to reduce the risk of this kind of accident on airplanes today. The Swissair Flight 111 accident was also a very complicated one. Much research was done on the subject of flammability of materials, particularly the materials used in the cabin for thermal acoustic protection. Flawed assumptions were a dominant characteristic of this accident. The Alaska Airlines Flight 261 accident involved a stabilizer control system that malfunctioned due to poor lubrication. Again, there were inconsistencies in the methods by which the aircraft was being maintained. The fourth accident that caused us to initiate this accident library was the American Eagle Flight 4184 accident in Indiana. There were flawed assumptions in the way that atmospheric icing would affect the airplane. The ice would actually accumulate on parts of the airfoil that had never been observed before. It was very unusual and, before this accident, it was a part of atmospheric icing that was not understood. The result was loss of flight control.

All these accidents and investigations resulted in a great deal of scientific research, testing, retesting, and evaluation. That research was documented, but was largely languishing in various archives and places not readily available to the general public.

Knowledge of these and other major accidents was basically being lost with the passage of time. The more years that passed, the more people forgot the causes, and the more new folks coming into this industry had no knowledge of these deficiencies and shortcomings at all, and we were seeing repeated accidents, which is unacceptable.

Awareness of information about previous accidents became key to understanding more recent accidents. The challenge that we undertook was to craft an information system that could take what could be well over a decade of work in doing the research and fixing things and make it freely and easily accessible to those who can benefit from it so we do not need to make these mistakes again.

More than a century ago, George Santayana of Harvard University wrote a paper titled, *Life of Reason: Reason in Common Sense*, saying that "those that cannot remember the past are condemned to repeat it."[1]

---

[1] Available at http://iat.iupui.edu/Santayana

It really is true. We will continue to repeat things unless we are mindful about what we need to be careful. A large transport airplane accident is an enormous human tragedy, but a second tragedy would be not to learn from it and then cause similar accidents.

Let me now talk about the barriers. Some other presenters talked about barriers to science and barriers to information flow. Aviation, particularly accidents, is driven by these very real and powerful barriers:

- **Fear of negative publicity**. There is probably nothing more negative than a large transport airplane accident. It is very sad for all involved.
- **Lengthy investigation.** Some investigations take more than a decade in order to get all the science and research together, and get a go-forward path that is solid.
- **Continual workforce turnover**. Many of the people that come into these accident investigations will be involved for a decade or so and then move on, and we replace them with new people and fail to build corporate memory of the problems.
- **The information technology (IT) tools**. If you go back 30 or 40 years, before the Internet and the computational systems, it was very hard to capture all these data.

We have developed a "Lessons Learned from Transport Airplane Accidents" library[2] that is organized, threat-based, and has search-and-sort capability. Its intent is to stop and reverse the loss of lessons so that others can benefit from what has taken 40 or 50 years to accumulate. Many of the speakers at this meeting, including myself, that came from out of town flew here on one of these airplanes. We take air travel almost for granted. The biggest risk is the car, not the airplane. We want to keep it that way and maintain a very safe aviation system.

Now let me talk about the portal itself. Right now there are 57 major accident modules on the site. We are working on more all the time. We have another five being crafted. It is relatively time-consuming to create this material and involves different stakeholders. Boeing, Airbus, General Electric, Pratt and Whitney, and the airline companies all have been very helpful. They realize that their work can benefit from getting this right.

The only information on the portal is information that is already publicly available. We are looking at maybe 10 to 15 years of hard work captured in a 15-minute read. Anybody who wants to know about accidents like these can get the big issues in about 15 minutes.

Regarding the organization of the library, when we first started this, it was very tempting to only look at issues from technical and scientific aspects, such as fire, structural issues, flight control issues, and things like that. Then we looked at several major accidents and asked ourselves: Are we really getting the maximum value of the material? If not, what are we missing? After looking at half a dozen big accidents that we had prototyped, we realized that we were missing the bigger issues beyond the technical and scientific ones; we were missing what may be an organizational problem or a human error aspect.

After the iterative process of developing the library, we agreed upon three perspectives of looking at accidents. First, we look at the accident from a perspective of what we call the airplane life cycle. This is the beginning, the operational, and the maintenance and repair perspective. Second, we examine the accident threat categories. In looking at accidents since the jet era began, it turns out that we can put them into 18 technical categories that are important. The third perspective is what we call the common themes. Every accident has a strong link to one of five common themes: flawed assumptions, human error,

---

[2] Available at http://accidents-ll.faa.gov/.

organizational lapses, preexisting failures, and unintended effects. Every miscue that has happened in aviation can be strongly linked to at least one of these. In some accidents we have found strong links to all five—not often, but there are some.

We have the site arranged so that you can search any one of these categories or themes. You can say, "I want to know about structural failures," and you can pull out the accidents in that category. You can also look at specific examples of accidents from an organizational standpoint. These are important not because they provide the most lessons, but they are certainly the most deadly accidents that have ever happened. For each accident there is a quick 60-second read of the general accident summary. The overview section gives you the nuts and bolts of the accident, the science, and the research that was done.

What we also feel is very important in this kind of material is the value of the Internet and IT tools—streaming video, animation, interviews, and things like that. Where animation has not been done by the companies involved, we have created animation with the help of staff at the National Aeronautics and Space Administration. It has been a very good partnership.

The portal has been in the public arena for two and a half years now. We are still building it. We add 5 to 10 accidents per year, and we will do that for at least 2 more years. Then we will transition into a sustaining mode where, as accidents happen, we will add them, but we will be finished with doing the historical mining and catching up.

14. Integrated Disaster Research: Issues Around Data

Jane E. Rovins
Integrated Research on Disaster Risk Programme, China

I am going to introduce a new program as well as talk about the data issues surrounding disasters, because this is a topic that the Integrated Research on Disaster Risk (IRDR) Programme has taken to heart. It is one of our three main projects. It comes up on a regular basis, and I spend as much time addressing disaster data questions as I do with risk-reduction questions.

So that you understand where I am coming from, I am a practitioner who happens to have academic credentials. I am a certified emergency manager. I spent about 10 years in the field getting mud on my boots before I came into research.

The main issues that are faced by the IRDR Programme in the disaster arena are globalization, population growth, widespread poverty, and climate change. Urban areas face unique challenges in these issues, so we have initiated several city-at-risk projects. The Coastal Cities at Risk project is a partnership between various countries and is funded by Canada. Another project is the Cities at Risk project, which is a partnership between IRDR, the Global Change System for Analysis, Research, and Training, and the East-West Center in Hawaii. These projects are geared toward the specialized needs of densely populated, rapidly developing cities.

In response to these issues, the International Council for Science (ICSU) decided to ask the following question: Why, despite the advances in science on hazards and disasters, do our losses continue to increase? They created the IRDR Science Plan to try to answer that question.[3]

In addressing natural and human-induced environmental hazards, we are taking an integrated approach to research through an international, multidisciplinary (natural, health, engineering, and social sciences, including socioeconomic analysis), collaborative research program. This, however, gets challenging. As you know, trying to bring even two or three of the natural science disciplines under one program can be difficult. IRDR is bringing all those disciplines in with the social sciences as well.

What exactly is our scope at IRDR? This is fairly straightforward, for those of you that have looked at natural hazards or disasters:

- Geophysical and hydrometeorological trigger events.
- Earthquakes, tsunamis, volcanoes, floods, storms (hurricanes, cyclones, typhoons), heat waves, droughts, wildfires, landslides, coastal erosion, climate change.
- Space weather and impact by near-Earth objects.
- Effects of human activities on creating or enhancing disasters, including land-use practices.

What we are not looking at are technological disasters and warfare. What do I mean by technological disasters? Take the Japanese earthquake and tsunami, for example. There were two natural disasters that created a technological problem with nuclear fallout. We would look at that in relationship to the earthquake and the tsunami. Had it been just a nuclear release caused by a failure at the plant, we probably would not have looked at it, but because it was created by natural hazards, we are incorporating it into our studies of the event.

---

[3] The Science Plan can be accessed at http://www.irdrinternational.org/.

The IRDR Science Plan, has three key objectives. All of these are very data intensive, as you will see. The first is the characterization of hazards, vulnerability, and risk. This includes the following:

- Identifying hazards and vulnerabilities leading to risks.
- Forecasting hazards and assessing risks.
- Dynamic modeling of risk.

This is mainly focused on identifying hazards and vulnerabilities, especially in developing countries. They do not necessarily have the means to identify and understand what their risk levels are (e.g., the forecasting of these hazards and assessing risk). The second objective is effective decision making in the context of complex and changing risk. This includes the following:

- Identifying relevant decision-making systems and their interactions.
- Understanding decision making in the context of environmental hazards.
- Improving the quality of decision-making practice.

The third objective has to do with how we ultimately reduce the risk and curb the losses through knowledge-based actions. In the disaster arena, there is much anecdotal information. That is historically on what the disaster field has based its decisions. Science is not good at getting information down to the practitioners. The practitioners do not understand a lot of the science, because it is presented in highly complicated expositions, using very statistics-heavy methods. It needs to be translated in such a way that they can actually understand it.

In this process, some of the things that we are trying to develop are long-term databases. Disasters have very short memories. Within a year or two of a major event, it is forgotten. Even in Florida, where they get hit all the time, they just forget.

How are we going to actually do this? The three crosscutting themes that we are implementing are as follows:

- Capacity building. Research is great, but if you do not train people how to use it, it does not do any good. We are partnering with several capacity-building organizations, in the United States and globally, to develop this.
- Looking at case studies and demonstration projects. People learn by seeing what has already happened.
- Addressing assessment, data management, and monitoring.

To illustrate the difficulties in achieving this kind of integration, we can think of the parable of the blind men and an elephant, where six blind men touch different parts of an elephant in a collective attempt to discern its true nature. The one who touched the elephant's ears decided that an elephant must be like a fan. The one who touched its trunk determined that an elephant is like a snake, and similar conclusions were reached by those who felt horns like spears, skin like a wall, legs like tree trunks, and a tail like a rope. Similarly, with so many scientists from diverse fields "touching" disasters, it is a monumental task to integrate their respective works into a comprehensive, cohesive understanding of the nature of hazards and risk. The researchers will examine wind speed, wave action, rainfall, ground motion, and the like. They all have their own unique data issues and their own unique data links to social science aspects, such as religion, ethnicity, age, and gender, for example.

Ultimately, though, what we need to be doing is talking about risk reduction in the collective sense, to get an accurate picture of what an elephant looks like. When you try to integrate census data from the social science side, which is collected differently and measured differently in every country around the world, with the natural sciences and how these issues are all measured, you come up with very different results. What IRDR is trying to do in the disaster arena is get researchers to look at all of the topics in a common platform and a common way. It is also very important that the private sector be brought into what we are doing.

Then there are problems with the quality and quantity of data on the impacts of disasters. This comes from several factors. Natural hazard management, right or wrong philosophically, is tied very closely to homeland security. When you start looking at disaster data and public infrastructure (e.g., police stations, fire stations, hospitals, power grids, water systems), the homeland security experts do not want to talk about it, because they are afraid that it will show vulnerabilities in the system. They would say, "We cannot talk about the water system, because then somebody would know how to poison the water system in a major city." I can understand that logic, but when you look at a city like Beijing, with 22 million people, whether or not that water system continues to function in an earthquake is really important.

There is also the issue of the quantity of data. There are massive amounts of satellite data coming in from all over the world for a disaster, but if we look at information on disaster losses related to injury and death, there is actually very little information because there is no common agreement within the medical field of what defines death by disaster. Disaster is not a cause of death on a death certificate. In the recent tsunami in Japan, for example, people were washed out to sea—is that an immediate impact? Or what about people in Florida who died of carbon monoxide poisoning several days after the disaster because they ran generators in their garage? Is that still death by disaster, or is that just accidental? We have not figured that out yet, and this creates questions related to the quality of data and how they are being analyzed.

There are many other biases affecting data quality, such as those in the databases regarding various losses. Most data related to disasters look at post-disaster responses and the losses that have accumulated. We also have hazard biases, since hazards are defined differently in different parts of the world. Further still, there are temporal biases. How are hazards compared over time? Can we compare a disaster from the 1960s and 1970s to a disaster that just occurred today? Realistically, it is much more difficult, partially because of technology. The technology was not available then to measure things the way we do now. Other biases include the following:

Threshold biases. This becomes a bigger issue, because many the databases will, for example, not count an event as a disaster unless a certain number of people were affected. For example, a tornado this morning ripped through Greenburg, PA. That community had less than 10,000 people, but the event was not any less significant. For the 11 families affected, it was as devastating as Hurricane Katrina was to the Gulf Coast. This becomes a threshold bias, because every database has a different cutoff.

Accounting biases. This is really straightforward. How are they measuring the economic effects? What losses are included and how is that done? Even within the U.S. Federal Emergency Management Agency, they are not consistent. There is no consistency across the field or within individual countries.

Geographic bias. This becomes really interesting when you start talking about changing borders. An example that occurred recently was an earthquake on the China-Myanmar border. How China counts the damages and how Myanmar counts them can be very different. How do you look at the earthquake collectively when you only get half the information?

If you are interested in these issues, I encourage you to look at a recent article titled "When do losses count? Six fallacies of natural hazard loss data"[4] that looks directly at natural hazard loss data.

So what is needed? For one, users need to be educated about data biases and issues of social loss data. This is especially important at the policy-makers' level. Within the social sciences sphere, this becomes an even bigger issue when you start looking at thresholds for age, income levels, and so on.

Making things comparable and accessible for human disaster loss data to support research and policy is another requirement. For example, the floods in Pakistan and China that occurred last year look very similar on many levels. The problem is, we cannot go in and scientifically compare them because the data were collected so differently for the two events.

Furthermore, existing databases need to be identified. We know what is out there globally, but what about the national and regional datasets? In Latin America, for example, they deal a lot on a regional level. What do the databases look like there versus here in the United States, in Canada, or in Europe? A lot of the data have global and regional trends. How do we make the sub-national databases accessible so that policy makers can look at them? We are working very closely with the United Nations International Strategy for Disaster Reduction (UNISDR) to try to make a lot of this information readily available to the policy makers so they can apply it in their areas.

The following are brief introductions to two activities already underway.

- The Risk Interpretation and Action project is looking at why people think they are at risk (or are not) and how they react. It is also taking into account the likelihood of the magnitude of events. Moreover, it is bringing together the physical sciences with the social sciences to look at the resiliency of physical infrastructure.
- Our job in the Forensic Investigations of Disasters (FORIN) project is to trace back through all the articles, data, and other research to figure out what happened leading up to and ultimately causing the disaster. What was the underlying cause for the event? It goes far beyond looking at why the communities are socially vulnerable. In Japan, for example, why was a nuclear reactor put in a tsunami zone? There must be an underlying reason somewhere back in history. This will be a series of case studies using a common template and methodology. The Japanese have already picked this up and will be looking at it in relationship to a previous earthquake, as well as the latest series of events.

How did IRDR get to this point, and who are our partners? We are cosponsored by ICSU, the UNISDR, and the International Social Science Council. Our partners are important, because this is about implementation. They are the national and international science institutes, national and international development agencies, and funding bodies. We are working with organizations such as the Department for International Development, which is the United Kingdom's overseas development agency, and the U.S. Agency for International Development, looking at the majority of disasters that happen in developing countries. We have six national committees around the world. These take the IRDR principles and apply them in a national context. We also have the IRDR Centres of Excellence, the first being in Taipei. They will be doing research (national and international) as well as capacity-building programs, and they have funding for young scientists.

One of the things that we are not doing is trying to duplicate existing initiatives. We are trying to identify the gaps, fill the gaps, and partner with agencies that are already out there doing it. We provide them with expertise and solid scientific research.

---

[4] M. Gall, K. A. Borden, S. L. Cutter, 2009. When do losses count? Six fallacies of natural hazard loss data. *BAMS* 90 (6): 799–809.

We strive to enhance the capacity of the world to address these issues and make informed decisions. We are trying to get stakeholders to shift their focus from response and recovery to mitigation and prevention. How remarkable would it be if there were a major earthquake and we did not have to spend the next 10 years rebuilding? What we are trying to do also is to learn from the data—not only the mistakes and the lessons learned, but also the best practices and innovative approaches.

15. Understanding Brazilian Biodiversity: Examples Where More Data Sharing Makes the Difference

Vanderlei Canhos
Reference Center on Environmental Information (CRIA), Brazil

Increasing consumption of natural resources is leading to a high rate of environmental degradation and biodiversity loss. In the last decade, large forest areas of the Brazilian Amazon Basin have been destroyed, including national park areas classified as high-priority areas for biodiversity conservation. To halt this destruction, real-time monitoring systems to support law enforcement are needed. Prompt and easy access to scientific and socioenvironmental data are critical for the development of real-time monitoring systems, law enforcement measures, and informed decision making. To promote conservation and sustainable use of biological resources, we need to improve access to quality and fit-for-use data, from molecules to ecosystems, deploying new tools that allow the real-time integration of different types of information, including legal regulations, biological data, and socioeconomic and environmental data.

Biological collections are repositories of reference materials and, as data centers, they provide the information necessary to understand the biocomplex interactions of organisms in ecosystems. Biological collections worldwide house an estimated 2.5 to 3 billion biodiversity objects. Of those, about 40 million samples are housed in Brazilian collections. Therefore, to improve knowledge of Brazilian biodiversity, we need to unlock and share the data associated with these Brazilian samples. The most efficient way to address this challenge is to develop a long-term, large-scale cooperative program focused on data sharing and data repatriation, with the deployment of state of-the-art information and communication technology.

The interoperability of e-infrastructures needs to be improved to facilitate the easy, open, and free access to biological data, tools, and Web services. This is the key to innovation in the life sciences, and is a critical factor in the design of support systems for informed decision making. Both the emergence of state-of-the-art computational tools and workflows to check and improve data quality and with the availability of Web services and tools that allow the visualization of models and data on a spatial scale are opening new avenues to improve our knowledge on biodiversity.

The 10-year existence of the speciesLink network[5] demonstrated that the key success factors for network consolidation were the adoption of practical solutions to address the lack of adequate institutional infrastructure and the change in culture of sharing data. The network was developed as an inclusive and collaborative bottom-up effort, based on low-cost equipment and free or shareware software. The project implementation started in November 2001, and the network was launched in October 2002, releasing 5,000 records online. Three years after the network launch, approximately 700,000 records were available online.

In the early days of the project, many organizations had serious concerns about releasing and sharing data, but this problem was overcome as the benefits of collaborative networking became clear. At the end of 2010, the network reached 4 million records from nearly 200 collections, double the 2 million records expected to be reached in the initial growth rate.

The key factor in the consolidation of the speciesLink network was the adoption of an architecture (Figure 15-1) that allows the harvesting of data from all types of collections, from small laboratories to large institutional repositories, including collections with substandard computing facilities and poor Internet connectivity. The development of mechanisms to mirror the data in regional cache nodes, associated with mechanisms to transfer and store the data in a centralized database, allows the deployment of tools and

---

[5] Available at http://splink.cria.org.br.

services to flag potential errors (such as misspelled names and other outliers) and the provision of tools to improve data quality to the data providers and ensuring full control of the data served.

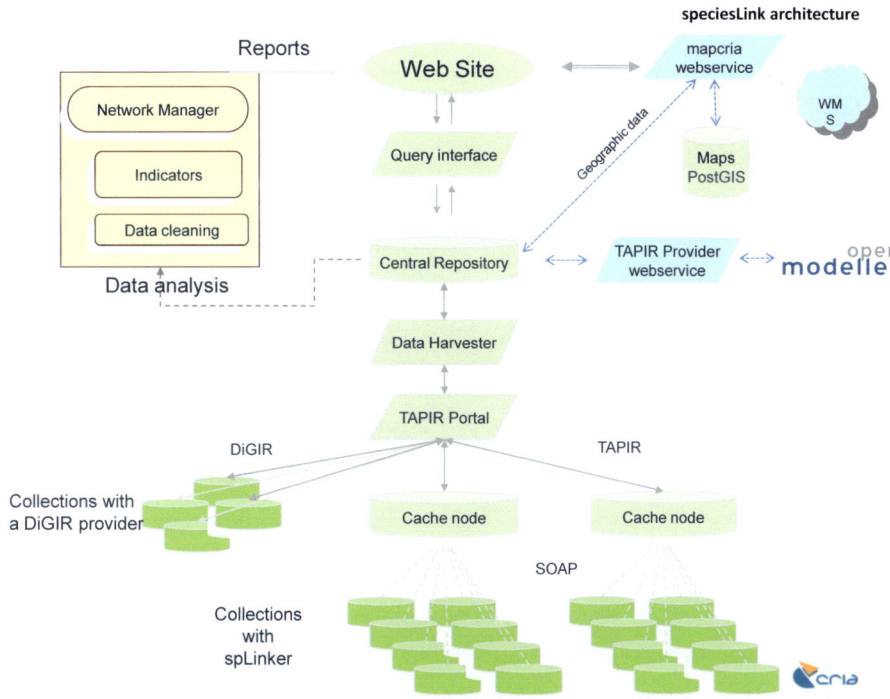

FIGURE 15-1 *speciesLink network architecture*
Source: From the speaker's presentation at the symposium.

Speakers in this symposium from Tanzania and South Africa emphasized that a major problem to be addressed is the lack of confidence of data providers to release and share data openly. This problem has been addressed by the speciesLink network through the development of mechanisms that allow the mapping and filtering of sensitive data, and facilitating the blocking of records that are not to be shared. When the user carries out the search, the system indicates that collection has the data, but data can be blocked. If needed, users can contact the database curator to check the restrictions to get the data under special circumstances.

The system harvests the data from cache nodes and stores the data in the central depository. This dataset is used for the network management and the development of indicators of the network evolution. The daily data-cleaning reports are available online not only to data providers but also to the users. This provides an opportunity for the users to check if the data have the required quality and are fit for use. A set of tools and algorithms are available to develop ecological niche models within the openModeller framework.

The 10-year development of the Reference Center on Environmental Information's (CRIA) biodiversity e-infrastructure is the basis for the effort to integrate the Brazilian and European e-infrastructures through the implementation of the "EU-Brazil Open Data and Cloud Computing e-Infrastructure[6]". This is a scoping project cofunded by the European Commission and the National Council for Scientific and Technological Development in Brazil.

---

[6] Available at http://www.eubrazilopenbio.eu/.

One important feature of the CRIA's system is the deployment of a statistics package to monitor usage. Figure 15-2 shows the continuous increase throughout the years, in which the green bar indicates the consolidated usage of all systems available at CRIA. The record is 4.2 million visits in 2010.

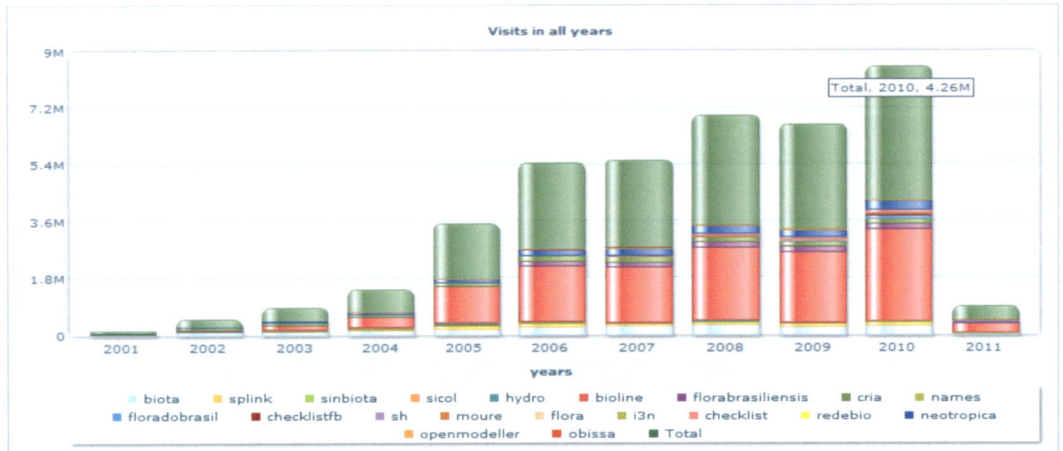

FIGURE 15-2 *Visits to CRIA's online systems in 2010*
Source: From the speaker's presentation at the symposium.

Considering that the bulk of information available at CRIA's systems is primary data and taxonomic information, the figures are impressive, mainly because more than 80 percent of users are from Brazil. This is the result of investments made by the Brazilian government in improving Internet access. The numbers show the increasing demand for access to scientific biodiversity data and information.

A challenge that is currently being addressed by the CRIA team is the development of mechanisms to share biodiversity information with different user communities, meeting the needs of scientists, policy makers, and common citizens. The way that the data from images, maps, species, and specimen is transparently integrated and shown to the users is key to meet their needs. New Web services are being developed to show high-resolution images integrated with taxonomic information. This is the core of the implementation of the ongoing, long-term, large-scale repatriation program of data from international collections.

In a cooperative effort involving the National Museum of Natural History (MNHN) in Paris and the São Paulo Botanical Garden, CRIA developed the "August de Saint-Hilaire Virtual Herbarium-HVSH[7]". This site integrates bibliographic data with higher-resolution images of maps, field notebooks, and samples of herbarium specimens collected during Saint-Hilaire's travels in Brazil between 1816 and 1822. This material is stored at the MNHN in Paris and the Clermont-Ferrand Herbarium in France. The information system is the prototype that will be adopted to integrate high-resolution images of samples collected in Brazil. This is relevant, as the MNHN Herbarium is finalizing the processing of more than 6 million images of specimens collected worldwide. After processing the label information and improving the data quality by adding coordinates and revising the taxonomic information, the digital data will be integrated to the MNHN Sonnerat Information System.

Another successful example is the revision of the Brazilian List of Plant Species. To meet the first target of the Global Strategy for Plant Conservation approved at the Conference of Parties of the Convention on Biological Diversity (CBD) in 2002, all CBD member countries were supposed to produce an updated

---

[7] Available at http://hvsh.cria.org.br/.

working list of plants by 2010. In 2008, after defining the strategy to revise the Brazilian List of Plant Species, the Brazilian government contracted with CRIA to develop the information system to support the collaborative network of experts. The coordination task was delegated to the Rio de Janeiro Botanical Garden. After merging more than 40 databases in different formats into a single one, the 400 specialists in the network revised the list. It included more than 40,000 valid names of plant species and synonyms, was published in May 2010, and was presented at the CBD Nagoya meeting in October 2010[8]. After the development of new functions, the system was reopened for the continuous review by experts.[9]

To conclude, we are living in a new information era where small nonprofit organizations like CRIA, in collaboration with local and international partners, can play an important role in the creation of virtual environments and e-infrastructures that integrate distributed data, information, and knowledge. These developments are opening new avenues for innovation in science and technology. In this respect the sharing of data, tools, and experiences is making the difference, and a big difference it is!

---

[8] Digital version available at http://floradobrasil.jbrj.gov.br/2010.
[9] The official 2011 updated list is available at http://floradobrasil.jbrj.gov.br/2011/.

16. Social Statistics as One of the Instruments of Strategic Management of Sustainable Development Processes: Compelling Examples

Victoria A. Bakhtina[10]
International Finance Corporation, United States

The focus of my presentation will be on social statistics as an instrument of strategic management for sustainable development processes. This topic is important because of the stronger interlinkages between social, economic, and ecological domains coming to the forefront of sustainable development progress, and a need to assess the impact of this progress on society.

The twenty-first century brings radical changes to determining the direction of economic progress, shifting the focus to solving the problems of innovative development, and transition to economies of knowledge based on intellectual resources, intellectual capital, science, and education. The ultimate goal is the improvement of people's lives and the expansion of people's choices and opportunities, together with sustainable, balanced, and harmonious development of society.

The modern vision of gradual sustainable development unifies three main components: economic, ecological, and social. The quality of the economic component is increasingly linked in people's minds with the concept of human development. Human or social development becomes the main purpose, and material means become a condition of this development.

Enormous data capabilities in conjunction with scientific and technological progress allow for a systemic approach to management and a comprehensive review and vision of complex information, leading to a better understanding of the major challenges facing humanity. It is key, therefore, to create an information-analytic base for innovative development using pioneering models and high-quality data. We can benchmark and look at the world as a whole, and we can add to these descriptions of reality and infuse more precision to the understanding of real situations.

Another approach is extracting data on risks, both general and specific, and drawing conclusions about management and prevention of risk situations. Statistical analysis is a necessary tool and condition for justifying the taking of strategic decisions.

In the 1990s the Human Development Report (HDR) of the United Nations demonstrated how to change the approach to development, focusing on people, and opening an era of new opportunities for the policy and research agenda. History has shown that a sole focus on economic growth does not necessarily result in the highest achievement in human development, and countries with the highest economic growth can lag in health, education, and other key areas. Having people as the source and the purpose of development makes progress equitable and creates an enabling environment for each member of the community to be a participant in change.

Initially the Human Development Index (HDI) encompassed only longevity, adult literacy, and income. In 2010, HDI was adjusted to reflect inequality and was complemented by the Multidimensional Poverty Index (MPI) and the Gender Inequality Index (GII), which illustrate countries' trends over the years and highlight important lessons for the future.

Assessments of "well-being" vary dramatically based on many aspects not covered by HDI: lack of self-respect and dignity, lack of access to assets and access to information, lack of time, insecurity, lack of

---

[10] The views expressed herein are those of the individual contributor and do not necessarily reflect the views of the IFC or its management.

PART THREE: COMPELLING BENEFITS 59

freedom of choice, helplessness, social exclusion, and isolation. This list is far from complete and can be expanded based on any specific context. These aspects are often interconnected and can lead to the disempowerment of people. Further research is needed to shed more light on these areas.

Social statistics complement and enrich economic and environmental measures, and allow us to analyze the efficiency of policies. It is equally important on a macro level when conceptually assessing global phenomena, and on a micro level, for example, utilizing household surveys, when assessing the qualitative nature of information and tailoring the policies to a certain context.

The Human Development Index is important for developed and developing countries, allowing us to track cumulative progress year after year. Measurements should be standardized as much as possible around the globe. Complemented by MPI and GII data, HDI adds various nuances to country performance relative to similar countries. The HDI shows that inequality adjustment can drastically shift the perception of the countries. For example, you can see from Figure 16-1 that there are significant losses in HDI due to inequality in various components.

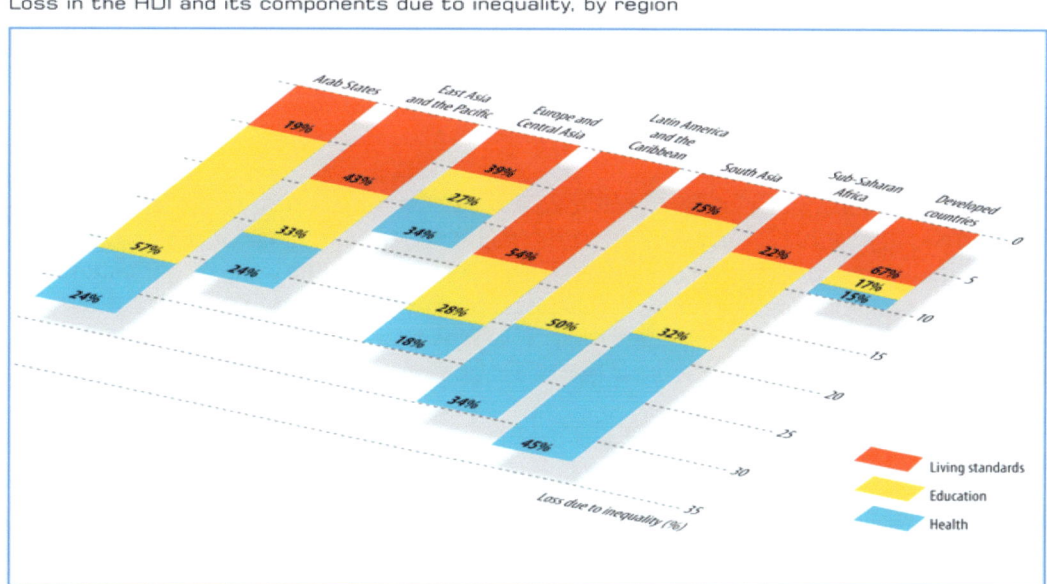

FIGURE 16-1 *HDI index adjustments due to inequality.*
Source: From the speaker's presentation at the symposium.

Figure 16-2 shows six African countries: Ghana, Botswana, Gabon, Malawi, Rwanda, and Kenya. These countries are dispersed around the continent, differ in access to natural resources, and have different political situations. There is a strong desire in these countries for reforms.

FIGURE 16-2 *Six African countries: Overall Satisfaction with Life*
Source: From the speaker's presentation at the symposium.

Rwanda has strong leadership, far-reaching reforms, and a roadmap outlining the transformation to a middle-income country by 2017. According to an International Monetary Fund report, reforms are on track and quantitative targets are being met. Malawi is a landlocked country and an agriculture-driven, smallholder-based economy. Ghana is one of the most robust democracies in Sub-Saharan Africa. Botswana is one of the most successful countries in Sub-Saharan Africa. Gabon is rated highest in human development in Sub-Saharan Africa. Kenya is one of the largest economies in East and Central Africa.

Figure 16-3 displays four of the six countries that are in the low human development quartile, according to the HDR. The main reasons are poverty, poor infrastructure, and limited access to health services and education.

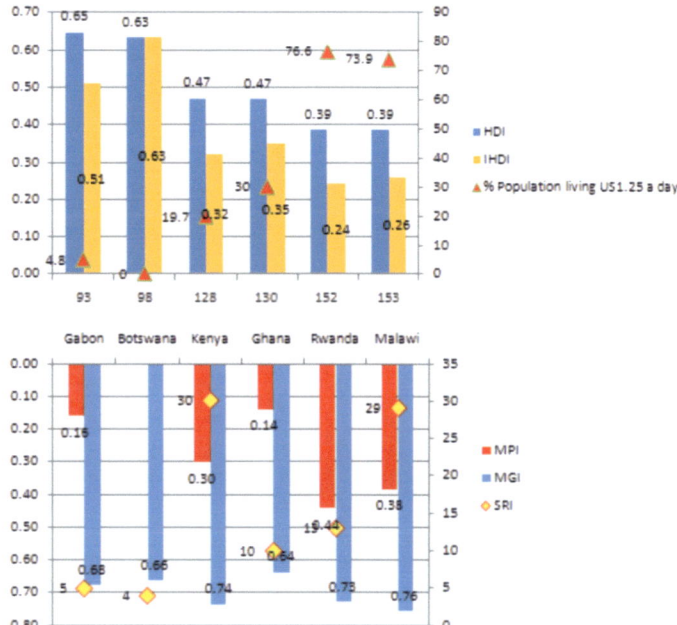

FIGURE 16-3 *Multidimensional Poverty Index of Six African Countries*
Data Source: HDR 2010; Figure is from the speaker's presentation at the symposium.

Gabon and Botswana have a medium level of human development, and they are almost identical, standing at 0.65 and 0.63 HDI, respectively. Kenya and Ghana are close, with 0.47 HDI. Rwanda and Malawi both have 0.39 HDI.

Let us expand the picture inferred from the Human Development Index further, and take into consideration inequalities in income, longevity, and education. According to Figure 16-3, Gabon moves to 0.5 HDI when we review inequality-adjusted HDI. After adjustment for inequality, Ghana comes first in the set of four countries from the low human development cluster. Malawi is a little higher than Rwanda.

Now, add an additional measure—a Multidimensional Poverty Index (see Figure 16-3). Based on this index, Ghana supersedes Gabon, appearing as the country with the lowest MPI of the six countries. In gender inequality, Rwanda exhibits the most equality, followed by Botswana and Gabon. Kenya has the same GII as Ghana, and Malawi shows the largest inequality.

An additional measure to be considered in the analysis is developmental, or the sustainability, risk index based on region-specific features. The sustainability risk index shows how vulnerable the countries are to the targeted key risks. According to the latest research (Bakhtina, 2010[11]), Botswana (rank 4) and Gabon (rank 5) falls into the cluster of least vulnerable countries, followed by Ghana (rank 10). Malawi (rank 29) falls into the medium-risk cluster, followed by Kenya (rank 30). Rwanda (rank 13) is the country least vulnerable to regional risks in the medium-risk cluster.

Now let us see how happy the people feel in the six countries under consideration. Overall satisfaction with life is highest in Malawi, followed by Ghana and Botswana, and those least satisfied with life are Kenyans. Females appear to be less happy. Is it a result of inequality? This presumption is not clear, as

---

[11] V. A. Bakhtina, 2010, Sub-Saharan Africa: Sustainability risk discussion (presented at CODATA 22 International Conference, Cape Town, October 24–27).

Malawi has the largest inequality among the six countries, but the highest life satisfaction (see Figure 16-4).

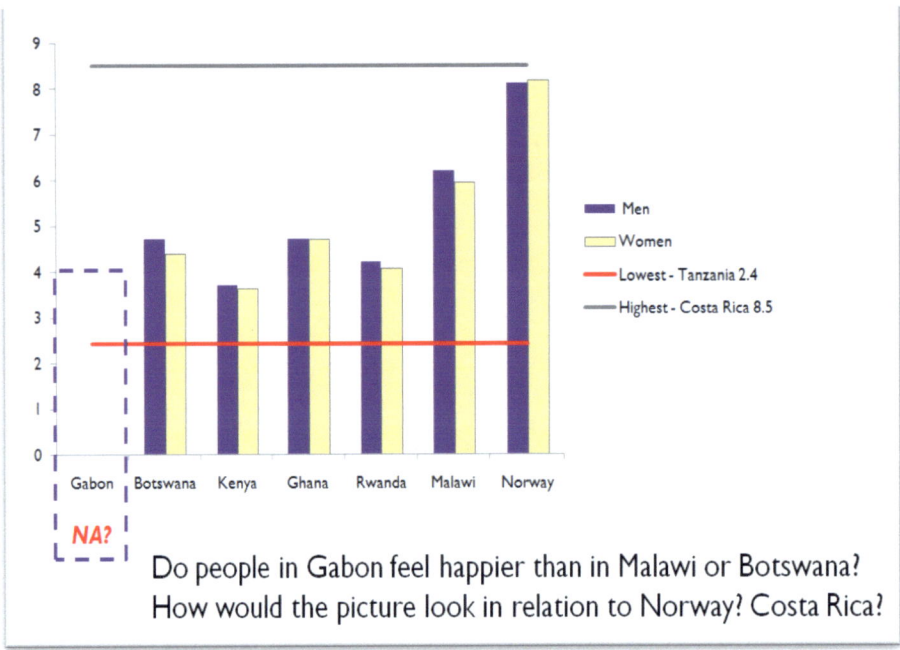

FIGURE 16-4 *Six African Countries–Overall Satisfaction with Life.*
Source: From the speaker's presentation at the symposium.
Credit: HDR 2010. Overall Satisfaction with Life is based on responses to a question about satisfaction with life in a Gallup World Poll

What conclusions can be made from these data? Would the data help us identify and flag key issues to address on the global scale? As we see from these statistics, life satisfaction has no strong relation to poverty and inequality.

When Malawi is compared with Norway, the country with highest HDI, the satisfaction-with-life index is 6 versus 8. (The lowest life-satisfaction level is observed in Tanzania at 2.4 and Togo at 2.6, the highest is in Costa Rica at 8.5.) We can also expand the research agenda: What policies would make people happier and inspired to act towards change and life-quality improvement?

Here, more information is needed: for example, measures of empowerment, unpaid women's work, and domestic violence. What factors increase happiness of the population?

According to the HDR, "human development is the expansion of people's freedoms to live long, healthy, and creative lives, to advance other goals they have reason to value, and to engage actively in shaping development equitably and sustainably on a shared planet. People are both the beneficiaries and drivers of human development as individuals and groups."

Let us turn now to some compelling examples showing how social statistics were used to drive policy decisions. In 2009 in Mexico, the Constitution and the General Law of Social Development were amended based on a multidimensional poverty measure reflecting various deprivations the households face (National Council for Evaluation of Social Policy). The country is mapped based on the level of deprivation in at least one of six dimensions: education, health care, social security, housing quality, basic household utilities, and access to food. The Mexican government uses the data to monitor the effectiveness of national social assistance programs.

The World Bank utilizes the Participatory Poverty Assessment (PPA) approach. The PPA allows the World Bank to incorporate views of the poor into the analysis of poverty, combine the results with other types of data, and communicate the findings to the policy makers, thus allowing the poor to influence policy. In numerous cases, PPA results contributed to a shift in World Bank lending programs. For example, in Ghana, PPAs contributed to a shift in the focus of reforms to rural infrastructure, quality and accessibility of health care, and education. In the 1990s, the World Bank focus in Nigeria was shifted to water and roads.

At the same time, PPAs were used in Thailand as a part of the Social Investment Project to increase the understanding of shifting patterns of vulnerability as the impact of the Asian crisis deepened, and to inform policy makers, strengthening the capacity of the country by consolidating various types of information. The Asian Development Bank also conducted similar assessments in Laos and the Philippines.

Social aspects are also key in disaster response and mitigation. Along with physical aspects, such as the probability of loss and risk, social aspects such as vulnerability and resilience should be considered. Fear, depression, despair, and post-traumatic stress can cause long-term consequences for the nation after a natural disaster. Apart from the assessment of hazards, probability, magnitude, and impact, a people-centered perspective is required to evaluate the susceptibility of the community to natural hazards. Interaction between hazard conditions and vulnerability conditions should be tracked dynamically (e.g., climate change impacts and economy impacts on migration of people to places susceptible to hazards like floods).

The World Health Organization points out that the concept of risk is associated with the perception of risk, and is a characteristic of society and culture. More effort is needed in community involvement in risk mapping and analysis, enhancing vulnerability assessments, understanding of risk perception, and the capacity to adjust.

The example of Mozambique shows how to link early warning of disaster and early action. After the flood in early 2000 left devastating consequences for the country, the social aspects of disaster mitigation were assessed and used to train the community to understand the risk and use warning information. The consequences of the 2007 and 2008 floods were significantly less severe.

Social factors contributing to cyclones' severity in Bangladesh are cultural. Male heads of the household historically did not want to move to shelters "unsuitable for females," with lack of privacy and poor sanitation. Families mostly adopted a wait-and-see strategy. Vulnerability was increased by late responses. Currently the issue is being addressed by educating the population.

The examples above show us how social statistics can serve as an instrument and basis for decision making and research on sustainable development processes. We are looking at the management process of decision making and utilizing this powerful instrument. We are showing how this instrument works on the global scale in the computation of indexes, which help us to understand objective trends. The focus is on the level at which the statistics are used. We reviewed three levels of application of social statistics: global, national, and local decision making. Many challenges remain, such as the sharing of data, the need for global statistical standards, data communication, intellectual property rights to the research results, and utilization of the data and research in policy decision making.

The benefits of sharing social data are many, and only a selection of them are given in this presentation. We should not forget that with benefits come responsibilities: To whom do we distribute the data? How do we protect the information against misinterpretation? How do we educate communities on how to use the data? How do we protect the data against terrorism? Who is accountable for quality, timeliness, and accessibility of data?

Let me conclude by reinforcing what I said earlier. We have tremendous opportunities. With the help of powerful technologies and new innovative models, we can promote and reinforce sustainable development. We can illustrate socioeconomic and ecological aspects of the development and address key risks. To validate, calibrate, and use the models, we need accurate and timely data and collective cooperation among statistical agencies and policy makers around the world. The assessment of the contribution of policies and financing mechanisms in the improvement of people's lives and expansion of their freedoms, cannot possibly be done without extensive shared information on social statistics.

References

Human Development Report (UN), 2010.

Robb, Caroline M., *Can the Poor Influence Policy? Participatory Poverty Assessment in the Developing World*, 2nd ed. (International Bank for Reconstruction and Development–The World Bank, 2002).

World Disasters Report 2009, Focus on Early Warning, Early Action (International Federation of Red Cross and Red Crescent Societies, 2009).

Bakhtina, Victoria A., Sub-Saharan Africa: Sustainability Risk Discussion (paper presented at the CODATA 22 International Conference, Cape Town, October 24–27, 2010).

17. Remote Sensing and In Situ Measurements in the Global Earth Observation System of Systems

Curtis Woodcock
Boston University, United States

What I would like to present today are my experiences related to the Group on Earth Observations (GEO), specifically the Forest Carbon Tracking (FCT) task. There are examples that relate to what we are interested in here, so I would like to share those.

The whole point of the FCT task has been to try to stay ahead of the problem of determining if and how developing countries could start to report on rates of deforestation and degradation. This is important so they could be eligible for compensation for reducing those rates, all in support of trying to reduce greenhouse gas emissions to the atmosphere to try to mitigate increasing carbon dioxide in the atmosphere. Then the main question is, can those countries report in a reasonable way? To do this accurately, there is a need for data.

The two primary kinds of data that are needed for this task are satellite observations and in situ forest data. I am mostly going to talk about the satellite observations. On the one hand, that is because if you think about data policy and data sharing, in situ observations are typically collected by people in their own backyard. On the other hand, satellite data are generally collected by somebody else. Certainly we ought to make the first step, that if we are going to collect data in other people's backyards, we ought to at least share the data with them, rather than necessarily force them to share the data they collect. If we could at least take that step, it would be a good step forward.

GEO has identified coordination among space agencies for image acquisitions as its first and top priority as it moves forward. Different satellite programs have collected data to be contributed to this task, including those in Table 17-1:

Table 17-1 *Image Acquisitions by Country*

ND Scene counts (vs submitted), 2010											
	Borneo	Brazil	Cameroon	Colombia	DRC	Guyana	Mexico	Peru	Sumatra	Tanzania	Tasmania	TOTAL
CBERS		2850 (14000)		1013		495		1864				6004 (14339)
LANDSAT	353		284	1097	731	225		1063	263	423	134	4354 (4366)
LANDSAT-INPE		685 (2130)		144		113		287				1189 (2191)
PALSAR	725	790 (3356)	150	803	1346	180	1004 (1286)	891	380	580	20	6798 (9101)
RADARSAT	35	38 (39)	37			27	123			70	80	410
RESOURCESAT		756 (1862)		120		117		282				1149 (1888)

Why I call it a first priority is that there has actually been some coordination of image acquisitions across different countries' space agencies. Those data were collected for this task and has been contributed to countries where we are trying to demonstrate these technologies. Nonetheless, there are still questions about data access, even though the data were collected specifically for this task. Free and open sharing of data remains very much an issue.

Why is this so hard? It is hard because satellite missions are expensive. They range from the hundreds of millions to billions of dollars. What makes it frustrating is that the revenues that countries get from selling data that come from these missions rarely offset anything close to a significant fraction of the cost of those missions. Just logically, it does not match up very well. I am going to try to convince you that charging for the data significantly hinders the use of the data.

The example I am going to use is the Landsat mission, which is the oldest of the land remote-sensing missions in the United States. There have been seven satellites in the Landsat program. Two are barely functioning still. The eighth is scheduled to be launched in December 2012. What we have is a dataset from 38 years of satellite observations that is a little more than 2.5 million images. An image covers 185 by 185 kilometers of the surface of the Earth. If you think about all the costs associated with building and launching the satellites, the data cost somewhere between $5 billion and $10 billion. It is an expensive dataset. We spent a lot of money on generating this dataset, and we would like to see it get used effectively in the future.

The good news is that starting in October 2008, the U.S. Geological Survey (USGS), the agency that distributes the Landsat data, stopped charging for the data and provided access on a no-cost basis to anyone in the world. The data usage has gone up by a factor of 100 since it became freely available. As a taxpayer, think about this. As shown in Figure 17-2, before the data became freely available, in the biggest year ever for sales of Landsat data, they sold about 25,000 images. The income that was coming in from selling images, on an annual basis, was somewhere between about $5 million and, at the very best, $10 million. That is one-tenth of 1 percent of the cost of the dataset, on an annual basis, that they were getting back from selling the data.

FIGURE 17-1 *Landsat Web-enabled Monthly Statistics*

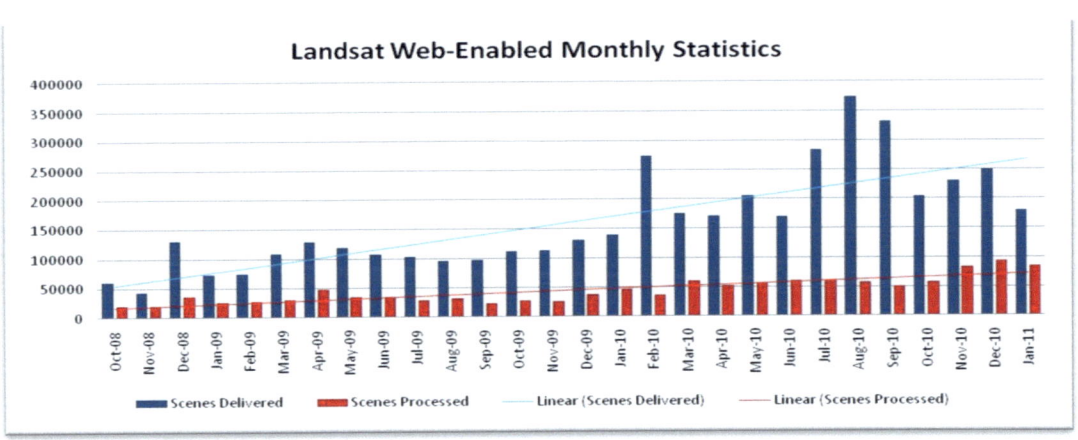

After all these years, we are finally getting our money's worth out of this dataset. This is the kind of effect that data policy can have on the ability to do research and applications.

The other thing that is interesting to note here is that I have been using Landsat data for my research over 35 years. All of a sudden, we are figuring out new and better ways to do a lot of things we never thought we could do in the past. It not only allows us to know what has happened in the past, but there are whole new categories of activities that are starting to show up, and it has been only 2 years since the distribution policy was changed. I think it is a transformative kind of experience for the use of satellite data.

Let me raise one sensitive issue about this data policy. Landsat is a U.S. satellite, but the United States has had international collaborators. The government sold licenses to receiving stations to get data over the years. There are about 2.5 million images archived in the United States and about 3 million images now in the international ground stations around the world. Although those other countries paid for the licenses to collect the data, the United States now is giving the data away freely. At the same time, the history of the surface of the Earth is embedded in these datasets. We cannot go back and recollect the data. It is a really valuable dataset showing part of the history of the planet's surface.

Here is where that effort stands. The USGS made an offer to the long list of countries that have data in their archives that if they would give us the data, we would reprocess them to the highest standards and return the data to them, and we would freely make available the data, if they want them. The United States has made that offer to these countries, but I think you can appreciate the delicate nature of this sort of discussion, when, after having asked people to pay for the data all these years ago, you are now saying that you want the data returned at no cost.

It really is an effort to bring into a standard state of affairs a very valuable dataset. A lot of this imagery is sitting around on media that is degrading. If we are going to do this right in the future, because we have not really done this correctly in the past, we need to think about the dimensions of international coordination and collaboration on earth observations. We now have dozens of countries running satellite observation missions, but in a very coordinated manner.

The first step is mission planning, so that there could actually be some compatibility between the datasets produced by these different organizations and countries. Then you have acquisition strategies. Can we get some coordination of acquisition strategies? For the first time, countries are actually beginning to collect data in concert with each other.

These satellite missions are expensive; there is no question about that. A big question then is risk mitigation. If multiple countries are going to put up comparable satellite missions, which is what is happening now, can we at least share risk with each other, so that if somebody's mission fails, the other will provide the data that would have otherwise been collected by the other satellite mission? This is starting to happen to some extent between the European Space Agency and the U.S. Geological Survey, with the Sentinel satellites and the Landsat program.

If we are really going to take advantage of having multiple satellite systems collecting similar data in space, it is equally important to get the data processed and distributed in consistent fashion, so that we can actually start to make use of them. It is difficult for individual investigators to try to take data from multiple satellite systems and combine them if they do not have any coordination at the outset. This is something that, at an organizational level, would really help.

My concluding thoughts are not too surprising. One is that we need satellite observations for many societal benefit areas, such as effective research and monitoring on climate and deforestation. We are in the infancy of trying to do any significant international coordination of these missions and collaboration, and there are many benefits to doing it that way. Free and open access to Earth observation, in some ways, is most important for developing countries, because they can least afford to pay for the data.

I also work in an organization called the Global Observation of Forest Cover and Global Observation of Land Dynamics. We have two sets of thematically oriented groups—the forest group and a land cover group. I cochair the latter. We also support capacity building through a set of regional networks. These are groups of regional scientists with common interests in Earth observation as used for forest monitoring and other such research and applications.

We have been running data-sharing workshops for the last couple of years, sponsoring people from regional networks to come to the USGS Earth Resources Observation Systems Data Center in South Dakota, and letting them take away all the data they want. This removes many of the technology barriers.

They go home with hard disks full of satellite images that they can use and share in the region. It becomes a focused network.

We are going to continue to do this for at least the next 2 or 3 years. There was one workshop per year for each of the last 2 years, and we are going to do two more this year and in the next 2 years. If anybody wants to try to get people into one of these workshops, or is interested in a particular region or getting access to the data that has already been taken back to some of these regions, I would be happy to help connect people and try to coordinate those activities.

PART THREE: COMPELLING BENEFITS 69

18. DISCUSSION BY THE WORKSHOP PARTICIPANTS

PARTICIPANT: This question is to Dan Cheney. Have you and the Federal Aviation Administration taken the next step? Now that you have all this wonderful scientific and technological experience data, have you started thinking about how you can exploit those data to make the entire air system both safer and more efficient?

MR. CHENEY: That really is the ultimate plan. The first step was to accumulate the knowledge. We are at the front end of taking the final step, and that is integrating the knowledge into our processes for maintaining and improving aviation safety. This initiative is only two and a half years old, so we are still in the middle stages of catching up on the history of aviation. We are laying the groundwork now for what to do with it. We know we have some gaps in knowledge in various aspects of aviation. So the question is, how do we use this to fill those gaps? We are not there yet. There is more work to do.

PARTICIPANT: I am thinking particularly about the accidents that did not happen. There is a huge amount of learning there that goes on inside the airline industry all the time, when these things are adequately reported. Sometimes they are not. There are malpractices with loading, with crew behavior in the cockpit, and so on. There are seriously overweight takeoffs and landings. We, as the general public, do not hear about those at all. There is an enormous amount of learning embedded in that.

MR. CHENEY: The aviation industry has benefited from a level of safety margins that exist in every aspect of safety. We have achieved internationally an unthinkable safety record when you look back at where we were 40 years ago and 30 years ago, and even 20 years ago. But it is because of the safety margins that are there. When we do have the overweight takeoff, when we do have the pilot that forgets to deploy a system, or a fire occurs, there are margins upon margins that result in that airplane not having an accident.

What we believe is the value in understanding the causes of yesterday's accidents is to recognize that when you do load an airplane over gross or you do have a tired crew that forgets things, you effectively take away one of the five levels of margin or one of the three levels of margin for that flight. That in itself may not be catastrophic, but now you have put that plane in a level of vulnerability in which it would not otherwise have been in. The accident is nearly always because you have eaten up four or five levels, and you have nothing left. The close calls, the accidents that did not happen—and they are certainly going on every day—are mirror images of what has happened in the last 50 years since the jet era. Whether it is an A380 or a 787 of tomorrow, those margins are the margins.

I think our big return will be to have tomorrow's decision makers and operators understand the importance of checklists, the importance of getting it right from a maintenance perspective, because getting it wrong takes the margin away on one level. Now we have a two-legged stool. One more mistake, and eventually we have an accident. The reason we are enjoying the safety we have is that it is a robust, margin-full industry. We are going to fight any threats of giving up any of the margins.

PARTICIPANT: One of the themes I got from several presentations is that a traditional problem for these kinds of data efforts is that data was locked up in disciplines, in silos. A lot of the presentations talked about how, especially for applied problems, you were able to overcome those traditional silos and put data together in new ways. The thing I am wondering about is, now that we have new problem-oriented silos, it seems there is always a danger of repeating history and having these new problem areas become silos on their own. I am just curious if you have thoughts on how to avoid that. To what extent in each of your areas are you really paying attention to other people's standards and looking at interoperability between biodiversity and disasters, which may not have an obvious connection now, but in fact, because of deforestation and forest cover loss, may have a connection to land use and disaster loss? To what extent do you think people are thinking, especially in developing countries, about not creating new barriers after having overcome the old barriers?

DR. CANHOS: I think that you raise a very critical point about standards and protocols for interoperability of databanks and data systems. When you look back at biological data, there were no standards and protocols for biodiversity data, although several organizations have worked in this area in recent years, such as the Global Biodiversity Information Facility (GBIF) in close collaboration with a Committee on Data for Science and Technology (CODATA) committee for developments and improvements in this area.

We do have big issues with silos. You talk here about different people working in different data areas such as biological data, species-oriented biological data and integration at a molecular level, genome data, and then integration with the information derived from satellite data, like land coverage. I think for local development, it is extremely important from the beginning to look at how the internationally agreed standards and protocols are developed. For the speciesLink network, when we started 10 years ago, our first approach was to see what was happening in the taxonomic databases working group. Also, after looking at what they were doing regarding biodiversity information, we decided to work in close collaboration with the Taxonomic Data Working Group (TDWG) and GBIF.

In conclusion, I think here we need the international organizations. We need CODATA and the International Council for Science (ICSU). All those efforts, like TDWG, are voluntary efforts. It is extremely difficult to develop standards and protocols on a voluntary basis. I think the major development was the support of GBIF to TDWG. This happened with $1.5 million that came from the Gordon and Betty Moore Foundation. Again, we need more involvement of international organizations. I think ICSU has an important role to play in the definition and further development of the data-related standards and protocols.

PARTICIPANT: Dr. Canhos, are you concerned about the sustainability of your biodiversity database?

DR. CANHOS: I am very concerned about the sustainability of those efforts. For me, it is difficult to get funds from the Brazilian government to maintain the speciesLink network. As to the content, from the beginning of the development of the network, we told the collections, "You can come in or leave the network anytime you want." It is just like pressing a button to take all the data from your database from the network, but that did not happen. Today we have more than 200 collections. The cost is well distributed and includes the cost of development of the software and the tools. This was a well-funded project for 4 years. Now the cost is to upgrade all those tools and also to develop more tools. The more data use we get, the more new requests for new applications we receive. I think that is a challenge, but now we are in an age of integrating distributed intelligence and distributed data. That lowers the cost. That is public infrastructure. When you think about the industrialized countries, we are talking about huge legacy collections. Think about the Smithsonian Institution. They have more than 130 million biodiversity objects in the Smithsonian collections. The cost to digitize all this legacy information is very high. Finally, I think we need support from international organizations. We need ICSU and CODATA to continue advocating the importance of not only sharing the data but also finding means to gather this data and to treat the data so that it can be easily accessible by the whole community all over the world.

DR. BAKHTINA: Regarding involvement of international organizations, I want to add that the World Bank promotes democratization of development via open data access. Last week, a new initiative, Mapping for Results, was launched. This initiative allows civil societies, various organizations, and citizens to access large World Bank databases, to contribute to the database and to use the data. It also opens a direct dialogue with governments and citizens, in terms of transparency of government policy and delivery of public services.

Furthermore, it will promote transparency related to the development, and hopefully, create incentives for various types of organizations to improve data quality and increase data sharing.

PARTICIPANT: I am with the International Environmental Data Rescue Organization. We are a nonprofit organization whose mandate is to locate, rescue, and digitize every piece of historic environmental data we can find throughout the world. We have projects in 15 developing countries, and we have rescued probably 2 million to 30 million historic weather observations. One of the problems we have is that, for example, we have located about 30 million weather observations on microfiche. They were taken in about 1,000 observation sites throughout Africa. We are negotiating with the African Center for Meteorological Applications for Development in Niger to at least get the microfiche before they deteriorate to nothing. I am wondering, do any of your organizations have data in a format where you cannot share them? Right now 95 percent of our data are either on microfiche or on paper, and that is a very huge problem for us. I am wondering if anybody knows of any other organizations, other than our own, that actually seek out data on perishable media to rescue them before they are gone forever.

DR. WOODCOCK: The U.S. Geological Survey is doing that for Landsat data. They do get data on all kinds of customized software and media, where they have to go back and reconstruct and reconcile them. It is hard. I do not do that myself, but I have been convinced that it is a pretty big obstacle in many countries.

PARTICIPANT: I am not sure if I am the only librarian in the audience, but libraries are definitely aware of this. They have been looking at this for straight text documents, and now they are starting to look at it for both born-digital, which is also being lost at an alarming rate, and print objects. So there are institutions working on these problems, and the National Science Foundation is supporting some of that activity.

PARTICIPANT: There is also a new CODATA task group called Data at Risk, which is working to at least identify a broader range of scientific data, not just environmental. Another group is the Minnesota Population Center. There is a lot of recovery of old census data, going back and migrating data on old media and similar activities. There are different groups in different disciplines that are doing this.

PARTICIPANT: The National Oceanic and Atmospheric Administration also has a data-rescue activity. I imagine there are quite a few others. And the United Nations Educational, Scientific and Cultural Organization, I believe, has a digital heritage program.

PARTICIPANT: I have a question for Dan Cheney about the difficulty in getting access to the data that you have been talking about because of the sensitive nature of them. For the FAA data, it could be potentially sensitive for the companies that either built the planes or operate them, in legal exposure. I was wondering if that is mitigated by some kind of legislation that caps the exposure to lawsuits or if there is some kind of waiver of liability associated with the disclosure of the data.

MR. CHENEY: I do not think I mentioned this during the talk. There are four criteria that have to be met before we even begin to look at an accident: the official accident report is issued; the corrective action and accident results are finished; there has not been another accident or incident that would call into question the official accident findings; and litigation is finished. Only then, when all four of those are met, do we begin to work, and we only use publicly available information. Normally it is the information that was gathered during the course of the investigative process, and in the United States it is part of the public domain. We work with other countries' accident-investigating bodies to secure access to their information that was gathered in the course of the investigation. We do not do any additional investigation. It is only what is already lying around in archives in pieces that is being lost. The task is to put it together in one cohesive place, look at it, and put it in a structure that makes sense. As far as litigation, the opportunity to litigate has already come and gone. Could someone come back and say, "Well, we did not think about that. Let us review it again." They could. Our greater-good concern, however, is that the information is so important that it cannot be lost, and we have to take this step and make sure that erosion is stopped.

PARTICIPANT: Do you also get data from other countries?

MR. CHENEY: We do. Not all are equally sharing, but there is outreach. As we have done networking and have improved our networking process, the efficiency is increasing, with recognition that this is a product that will benefit world aviation, not just U.S. aviation. It is getting much better.

PARTICIPANT: Victoria Bakhtina, how difficult is it to get access to the data that support the statistics for the different countries? Are the data collected by the countries themselves and made available to the World Bank, or does the World Bank pay people to collect the data and work with the different ministries? It is not clear how the data are compiled, and whether it is comparable kind of coverage.

DR. BAKHTINA: In terms of access, with the launch of open data policy, it becomes very easy. Now the set of all development reports that include underlying data will be available. Just visit the World Bank website, and you can easily download and research the statistics from different countries. Moreover, you can visualize the data and map the country statistics to the World Bank projects information. I am also using the UN data which are also publicly available. There are various methods and strategies of collecting the data and multiple approaches can be applied depending on what is measured. The World Bank Data Group partners with other organizations on data collection and statistical capacity-building. I would recommend that you consult the World Bank web site for details. When using any publicly available data, it is important to understand what is behind the numbers, and depending on the end goal of your research, determine what coverage would be acceptable, and most importantly, always conduct the analysis within a specific context.

PART FOUR:
THE LIMITS AND BARRIERS TO DATA SHARING

19. Data Sharing: Limits and Barriers and Initiatives to Overcome Them – An Introduction

Roger Pfister
ICSU Committee on Freedom and Responsibility in the conduct of Science, France, and
Swiss Academies of Arts and Sciences, Switzerland

Discussing the limits and barriers to data sharing with developing countries is also about their access to information. This paper suggests that the situation is improving, albeit slowly. We will look at the developments in this area from a historical perspective. A review of the last few decades will allow seeing the current situation against that background.

Calls for a new information order
To locate the idea of information sharing in a historic perspective, we go back to the 1960s. Mention needs to be made of the Non-Aligned Movement (NAM) in this regard. Established in 1961, the mission of this political organization was to pursue an independent policy based on coexistence between the two power blocks, East and West, dominating the Cold War era. NAM was basically the device of the developing world, because it comprised primarily countries from Africa, Central and South America, and Asia.

Following political independence, especially on the African continent, during the first part of the 1960s, members of the Non-Aligned Movement strived for economic and cultural liberation from the North, which could be seen as synonymous to the West in those days. For that purpose it propagated two initiatives during the first half of the 1970s, namely the New International Economic Order and the New International Information Order. To gain international attention and support for them, and by means of political lobbying, the latter was taken into the United Nations Educational, Scientific and Cultural Organization (UNESCO). The reason for this move was that the number of developing countries had increased so much in that UN body that developing world demands could be pushed through rather easily. As a result, the Non-Aligned Movement succeeded at the UNESCO 19th General Conference in Nairobi in 1976. A resolution was adopted that called for the free flow of information. This was facilitated by the fact that the UNESCO Director-General at the time was Amadou-Mahtar M'Bow from Senegal. However, the two initiatives came to a standstill in the mid-1980s because political quarrels over the organization's role in this field limited UNESCO's possibilities. A crucial loss was when the United States, by far the largest sponsor, left the organization in 1984 over that issue.

Yet, the NAM's demand for a New International Information Order had a sound basis. The notion of developing countries having been left out of the worldwide flow of information was based on realities that we would like to illustrate with some statistical data from the period before e-mail and Internet emerged as tools of communication and tools for exchanging information. These were the days when newspapers, television, and radio were the principal means for spreading information.

Indicators of information sharing
Figures on the number of daily newspapers available in the different world regions reveal that Africa had the fewest between 1980 and 1994, with no increase in the period under consideration, and that all developing areas fared much behind the developed Europe and North America (Table 19-1).

TABLE 19-1 Daily newspapers per 1,000 people

	1980	1990	1994
N. America	247	231	209
Europe	188	220	244
Asia	87	106	116
S. America	88	95	98
Oceania	64	44	46
Africa	14	16	14

UN Statistical Yearbook; UNESCO Statistical Yearbook.

Another statistical figure indicating the underprivileged access of developing countries to information is the low number of television receivers per one thousand habitants. At the same time, it is significant to note the sixfold increase in numbers in the developing parts of the world between 1980 and 1997, as compared with only some 25 percent in the developed regions (Table 19-2).

TABLE 19-2 TV receivers per 1,000 people

	1980	1990	1997
Developing Countries	27	124	157
Developed Countries	424	492	548

UNESCO Statistical Yearbook

Radios have been an even more important source of information in the developing countries, and an almost 400 percent increase can be noted from 1980 to 1997, as compared with only 50 percent in the developed countries (Table 19-3).

TABLE 19-3 Radio receivers per 1,000 people

	1980	1990	1997
Developing Countries	398	895	1,124
Developed Countries	986	1,181	1,308

UNESCO Statistical Yearbook

All of the above figures indicate that the issue of information flow has been of great relevance to the developing countries, and that some development and progress can be discerned with them gaining increased access.

This is to put into perspective the situation concerning the access of these countries to information, which has become increasingly and more easily available since the advent of the Internet in the mid-1990s. Figures, once again, illustrate that developing countries are lagging behind. However, there has been a phenomenal increase in the number of people in those regions using the Internet in the years from 2000 to 2010 (Table 19-4). There is still a long way to go to reach dimensions such as in North America or Europe, but there is progress.

TABLE 19-4 Internet growth 2000–2010
Internet users

	Growth (%; 2000-2010)	Penetration (% Population)
N. America	146	77
Europe	352	58
Asia	622	22
S. America	1,033	35
Africa	2,357	11

http://www.internetworldstats.com/stats.htm

Initiatives of the International Council for Science (ICSU)
Against this background, the International Council for Science (ICSU) recognizes that access to information is crucial for both science and for a world where science is used for the benefit of society. A cornerstone of ICSU's mission, therefore, is to promote the universal and equitable access to data and information. Several ICSU initiatives sustain this policy approach and are now mentioned in the chronological order of their establishment.

The general objectives of the Committee on Data for Science and Technology (CODATA; established in 1966) are to improve the quality and accessibility of data, to facilitate international cooperation among those collecting, organizing, and using data, as well as to promote an increased awareness in the scientific and technical community of the importance of these activities. The International Network for the Availability of Scientific Publications (INASP) aims at improving access to scientific and scholarly information; fostering in-country, regional, and international cooperation and networking; and advising local organizations and funding agencies on ways to utilize information and publishing to achieve development goals. Most recently, the World Data System (WDS) was established in 2008 to *inter alia*, enable universal and equitable access to quality-assured scientific data, data services, products, and information. The WDS is being built on the foundation of two earlier international networks that were established in the context of the International Geophysical Year in 1958—the World Data Centers and the Federation of Astronomical and Geophysical Data Analysis Services.

Apart from these bodies with a specific remit for promoting data and information sharing, three regional offices—in Africa, Asia and the Pacific, and Latin America and the Caribbean—ensure that ICSU's strategy and activities, among them its approach to data and information sharing, are responsive to the needs of developing countries.

Finally, the ICSU Committee on Freedom and Responsibility in the conduct of Science (CFRS) is also concerned with these issues as part of its mission to promote the Principle of Universality of Science.

This principle is about developing a truly global scientific community on the basis of equity and nondiscrimination, which also comprises equal and nondiscriminatory access to data and information. For this reason, the committee, and several of its members, cosponsored this international symposium organized by the U.S. National Academies, which forms the basis for the present publication. To further raise awareness among the global scientific community for these concerns, the CFRS issued an Advisory Note on the matter following this scientific meeting[1]

---

[1] The CFRS Advisory Note is available in Appendix C and at http://www.icsu.org/publications/cfrs-statements.

20. Consideration of Barriers to Data Sharing

Elaine Collier
National Institutes of Health, United States

I will give a brief overview of the things we should think about when considering the barriers to data sharing. I will focus on questions related to data, although scientific information is equally important.

First, let us look at finding out about the existence of data. Many questions come to mind, such as the following:

- How do you know whether the data that you want exist or not, and how do you find out?
- Is the database discoverable by humans or machines, and human or machine readable and usable?
- Do you have to know a friend of a friend of a friend to get access to the data or information, or can you discover it on the Internet, or on other media?
- Once you discover that the data exist, can you discover what the characteristics of the data are, and if the data are usable in ways that you want to use them?
- Do you know what parameters there are among the data?
- What are the requirements for access to the database, and can you discover what those requirements are?

All of this should be easy. We should know whether the data exist or not, but whether we have access to them is a different question. Often their very existence is difficult to determine, particularly in scientific areas.

Some of the other issues relate to the actual characteristics of data, including semantics or meaning of the data, for example:

- What are the elements and what are the fields in the data, and what do they mean?
- What formats are they in so that you can actually use them?
- Are there readable words on a page or are the data in a data field in a relational database?
- Are the data on the Web in semantic Web technology?
- What kind of information and protocols are the data in?
- Is the database complete? Are you getting the raw data, aggregate data, or derived data?
- What is the history of the data? Who collected them? On what authority? Who curated the database, added to it, or annotated it?
- Does it link to other data or other information that is out there? Is it public or private information? And again, how do you get access to that data and information?

When we talk about semantics or data quality, we are referring to the actual meaning of the data. That has to do with the content. There are also some questions in this regard:

- What is the content information in the data or information that you have? What is it about? What does it mean?
- What is the context of that content? Is it seen from a perspective of a certain country or a profession?
- What are the temporal aspects? Are you getting the early information or the latest information?
- Do you have the whole picture of the time frame of the data?
- What is the granularity of the database? Do you know the details about the data or only high-level information?

- In what language are the data? This is a particularly interesting issue when it comes to sharing information across countries and the world.
- What is the durability of the information? Will it be there tomorrow or is it ethereal and passing?

Then there are the annotations of the data, and the properties or other context information about the data. In what framework are the data? What are the community needs and requirements about the data, in the collection of the data, in your analysis, and in your use? Are there shared standards for this information, so that you can actually mix them with other data that are similar from other places and other countries and other areas?

Some of the technical aspects relate not only to having standards for all these data quality measures, but also to being able to share the format so you can actually share the information electronically. Or if you are going to share it on paper in human-readable form, it again depends on what language it is in. It also depends on whether it is on a piece of paper or in a file that you can download. What is its availability? Is it persistent? If it is persistent, what version do you have of the data? Is it updated?

Are there requirements for reusing the data or repurposing them? When you do that, what are your requirements of the database? Do you have to clean it? Do you have to derive it into other formats and document what you are doing with it? Did other people who did that to the data before you document them so that you know what happened to the data?

Are the data coming from a repository, where the people you are getting them from are merely serving as the repository or are they actually the publishers of a database in the sense of publishing a dataset, not just publishing a paper? Again, is the data linked to other data, or is it able to be linked to other data? How do we preserve the data so that we actually have a history of what is going on, particularly as technology changes?

Some of the issues relate to policy and cultural issues. Some people who collect data are concerned about their misuse or their misinterpretation, and are reluctant to share them because they are afraid somebody will use them wrongly.

There are confidentiality and privacy concerns. Some relate to human data from clinical studies. Some of the concerns relate to health care data, while others are related to competitive advantage and proprietary information. There are also legal considerations across countries, which are very complex and relate to privacy issues.

Intellectual property is another interesting issue. One person's intellectual property and one country's intellectual property is not necessarily the same as another's.

What resources does it take to make your data available for sharing? Even if a country wanted to make data available, how would they do it and would they have the resources to make the data accessible? What are the resources needed to access other people's data? What kind of approval process do you need? What are the costs? The costs here are not only technical costs, but also the cost of policy agreements that may be required to use the data or to make them available.

Some of the competing interests arise as a result of different perspectives. One perspective is that of a researcher or a collector of the data. That can be a country, a utility, or a scientific researcher. That individual has certain interests in the data, but the institution, the company, or the country government where they work may have interests that may be different from the collector's. Second, there are national issues related to having competitive advantages. Third, there are certain advantages for cooperating and sharing data.

Next, there are the public and private issues. What are public data and what are private data? Could certain information that you consider public be considered private in some areas? Some issues relate to privacy and security. What is the impact of sharing the information on the individual? Is it an individual's information? Is it the individual who actually collected the data? What is the impact on the institution? Will the institution look bad if they share the data? If you have hospital data and infection rates and you share them, will that impact badly on one hospital because they have a higher infection rate than another, or is that related to the patients they see? Public health data are critical to share in order to prevent public health outbreaks, but such data also can affect the reputation of countries and institutions. Some of this is cultural. There are more open groups, people, and institutions that are sharing and being willing to risk more. How much of a database should we share? Should we share the raw or aggregate data? How do we make the data unidentifiable? Does that make them better or worse to share?

Clinical care and clinical research are particular issues related to privacy and security, and the latter are clearly very different across different countries. First, you have questions about the audience. One audience includes both the people who are actually sharing their data and information; the other is the people who are using those data. Then we consider cost. Is it the cost of the institution or the cost to the person? What is the cost versus the value of the information? Is it more valuable to spend the money to get the data or not? We also need to consider the effect of cost on the usability and the availability of the data. Free data may be very valuable or worthless. They could also be very expensive data to buy in cost or in effort and ultimately not be worth anything. How do we get these issues worked out?

21. Artificial Barriers to Data Sharing – Technical Aspects

Donald R. Riley
University of Maryland, United States

When I talk about technical issues, I speak as a mechanical engineering professor who taught for 22 years at the University of Minnesota. I grew up with what eventually became the Internet. I also grew up in the university environment and have remained inside it throughout my career. My view of the world is that we teach and do research that has real impact on society. To do that, we need certain kinds of tools and access to data, resources like cloud computing, and so on.

I also have been a chief information officer, meaning I had to provide and be responsible for infrastructure and tools and services in two major research universities. There was a need to develop advanced capabilities and create an infrastructure that had certain characteristics to serve those missions. This led to the creation of Internet2, which has become a global activity. I think it is fundamental and crucial. I am also now part-time chair of the Internet Educational Equal Access Foundation, which has a goal of trying to get universities and schools connected around the world.

What the statistics about regional usage of the Internet do not show is whether or not people have to drive or walk or bike 2 miles to an Internet café to get that access. Is the university connected, and at what speed? In Table 21-1 and Figure 21-2, you can look at some of the statistics, in aggregate, but numbers sometimes do not tell the whole story.

TABLE 21-1 *2010 Internet World Statistics*

Region	Population	Internet Users	P. R.	% Users
Asia	3,834,792,852	872,526,978	22.8 %	43.0 %
Europe	813,319,511	475,123,735	58.4 %	23.4 %
North America	344,124,450	271,330,900	78.8 %	13.4 %
Latam / Carib.	592,556,972	209,874,973	35.4 %	10.3 %
Africa	1,013,779,050	115,631,340	11.4 %	5.7 %
Middle East	212,336,924	63,708,386	30.0 %	3.1 %
Oceania, Aust.	34,700,201	21,272,470	61.3 %	1.1 %
Total World	6,845,609,960	2,029,468,782	29.6 %	100.0 %

Credit: Copyright © 2000 - 2012, Miniwatts Marketing Group. All rights reserved.

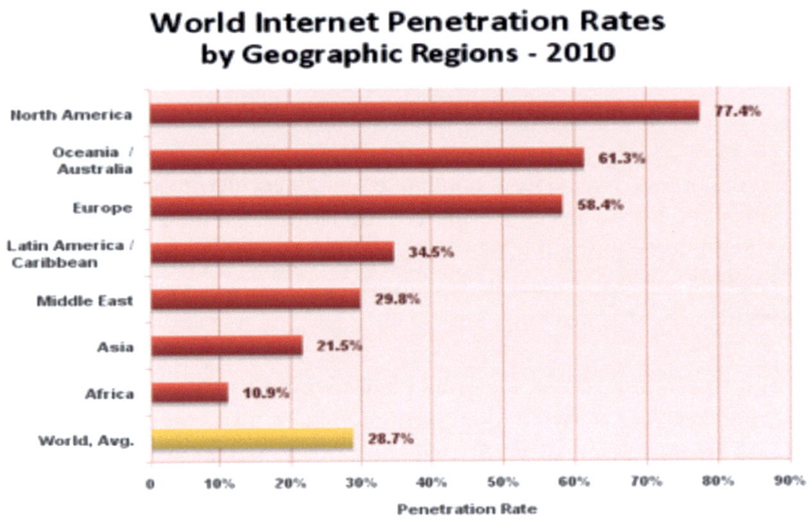

FIGURE 21-1 World Internet Penetration Rates by Geographical Region in 2010
Credit: Copyright © 2000 - 2012, Miniwatts Marketing Group. All rights reserved.

When we look at Africa, it seems like the situation is improving. People are really excited about cell phone penetration rates, because it generates some revenue and it fosters connectivity, but each device's speed and capability affects what you can actually do. Even these limited devices are expensive when your monthly income is less than $100 a month. We have to consider more deeply the quality, access, and affordability.

Is it about the Internet or is it about next-generation kinds of things that you can explore and develop and be part of the technology generation with new information and new tools? How do we collaborate beyond e-mail and tweets? From my perspective, I think universities are the cornerstone of technology development implementation. The biggest problem we have in Africa is that there are not enough people who know how to manage routers and other such things, and how to deal with bots and other malware.

The real message is that we need to look beyond just focusing on the Internet and whatever the telecommunications companies provide, and look at the future. The emphasis should be on performance: advanced capabilities instead of just more bandwidth. If we really want to have a society that is able to compete in the global information economy, then we have to focus on high performance and collaboration without barriers. We have to look beyond bandwidth. We have to look at how we support advanced collaboration, teaching, and research. We need to focus on quality and recognize it is not just about saying there are things you can do from an educational standpoint, but it is also fundamentally about creating that enabler for advanced economic development and sustainability.

The model that is evolving on the international scale is similar to our Internet*2*: the National Research and Education Network (NREN) that manages its own backbone and capabilities. In the United States, we have state-based regional optical networks that all tie into this big backbone, and with the Obama administration Broadband Technologies Opportunities Program, we are now talking about reaching out and creating "community anchor" networks in, for example, rural areas and inner cities and in other areas that need improved broadband access—that then tie into this new universal community-access public-purpose backbone.

My next message concerns NRENs. If you look at the map showing international collaborators in Figure 21-2, the NRENs that are mostly in the northern hemisphere and the more economically developed countries. There also are some light gray areas, mostly in the southern hemisphere, where there is no equivalent NREN in place yet. Most of those gaps are in developing countries. Fortunately, however, there are already some networks from the developing countries represented, so it is getting better.

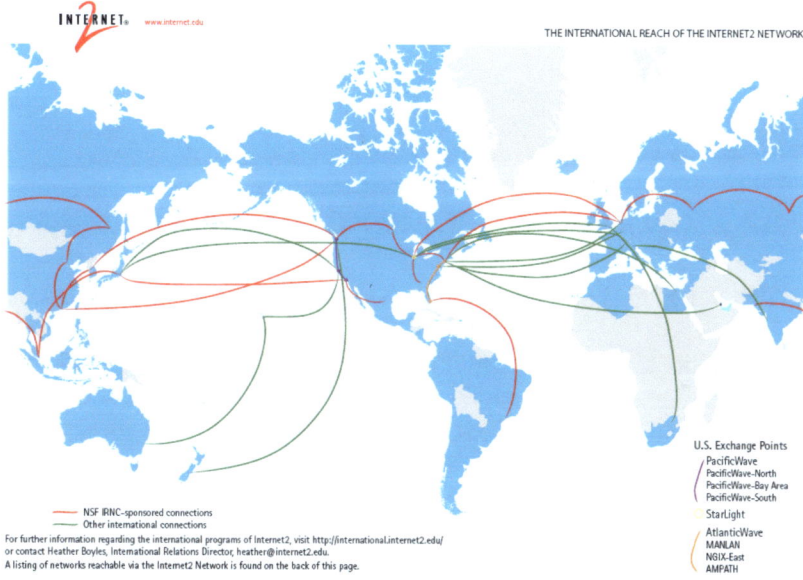

FIGURE 21-2 The International Reach of the Internet2 Network

If you want to know more about NRENs, the European Networking Research Organization does an annual compendium of the NRENs around the world. They also produce a study called *The Case for NRENS*[2], which looks at where NRENs exist and what their impact has been. A significant conclusion that they drew is that cutting-edge advanced capability networks provide services and encourage and enable technology spillover into the commercial sector. They also concluded that where NRENs do not exist, it hampers development and can exclude countries from achieving advances that could help their economic development.

If we look at the Global Lambda Integrated Facility, which is a global research platform for advanced applications, they have multiple 10-gigabyte-per-second links, a level of connectivity that gives it the capability of shared and collaborative research, connecting different parts of the world, with one such connection to Africa in South Africa. In African nations, and other developing countries, international connectivity is poor and expensive, because

- Internet cost is very high (Figure 21-3);
- Satellite access limits what can be undertaken, because of latencies and asymmetrical characteristics (it assumes Africa is a user of, not a generator of, new information); and
- There are significant barriers to access to information and resources, modern education, collaboration, research, and funding opportunities.

---

[2] Available at http://www.terena.org/publications/files/20090127-case-for-nrens.pdf

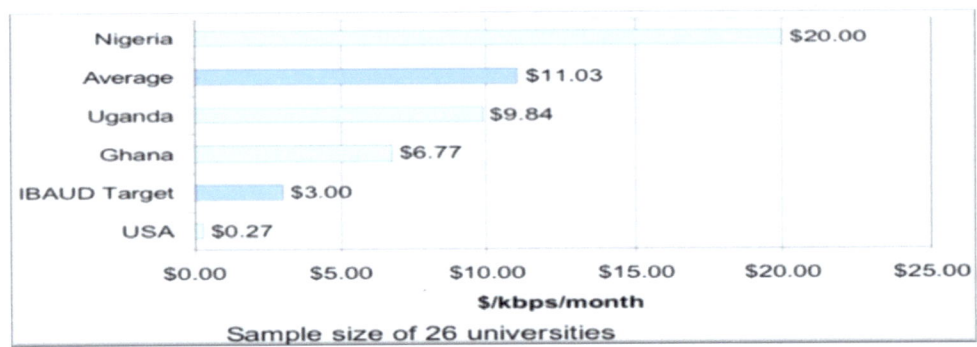

FIGURE 21-3 Sample bandwidth costs for African universities

Much of it is based on satellite service, although they are starting to build terrestrial fiber to interconnect to submarine cables. Submarine cables are not thought of frequently, but if we did not have submarine cables, it would be difficult to transmit data back and forth. There are cables coming to Africa, but the gap is still widening. There are studies you can look at that relate to technical aspects and capabilities, but the human infrastructure is not being developed, because they do not have access to these capabilities.

NRENs are developing in a lot of African countries, but they have a long way to go. Most of the new technologies are hard to deploy because of unreliable infrastructure, such as Skype calls, videoconferences, audio conferences, and the like. However, new submarine cables are being deployed around Africa in different places, and now there are fiber-optic cables on the ground to connect land-locked countries to the sea route and to the submarine cables. This connectivity is evolving slowly, however.

The UbuntuNet Alliance covers southern and eastern Africa, and is an alliance of NRENs trying to determine what their common future needs to be. They have lots of challenges, such as the telecommunication monopolies. Few of those countries have truly adopted competitive telecommunications practices.

I will now turn to some technical and managerial challenges. Working across all these countries, there are many issues to consider: technical aspects, the lack of human resources, and financial undersourcing, along with high cost, uneven development, and insufficient governmental support. Additionally, we are faced with the high costs of connectivity; network and equipment costs; inefficient use of established networks; an uneven development of technological infrastructure related to the different sectors; and insufficient governmental and administrative support for the development of information and communication technology infrastructure. Furthermore, the collaboration among research institutions in the region is not at the required level. There is a lack of skilled human resources and knowledge for its implementation.

The reason I am here is because Paul Uhlir and I worked together in different meetings. I cochaired the InterAcademy Panel on International Issues Task Group on Knowledge Infrastructure. Some of the things we looked at were related to data sharing and open repositories, and what we need to really promote at the Academies. Wherever possible, we also focused on government support for promoting the high-speed, specialized networks to connect universities. We wanted to emphasize that the topic does not necessarily have to be scientific. It could be liberal arts. It could be shared performances that need infrastructure dedicated to information flows and collaborations. What is really needed are clear policies, consistent regulations, and plans and development of the underlying factors. We included some recommendations that were specifically targeted toward governments, including things that they ought to do, and things that those of us in research and educational organizations can do as well.

22. Scientific Management and Cultural Aspects

David Carlson
University of Colorado, United States[3]

Suppose that based on everything we have heard for the last day and a half, we took the best information and the most advanced and thoughtful practices and we decided to conduct a big science program. Let us adhere strictly to free and open access, and do it in the age of cyberinfrastructure and global connections and aspire to set new standards for data access. What would happen? Well, the International Polar Year (IPY) did that. It was the largest program since International Geophysical Year in 1957–1958. Out of IPY came some interesting lessons.

These lessons have nothing to do with developing versus developed countries. Rather, they come out of the science culture itself. That is what I want to talk about today: data lessons from the International Polar Year.

The two big organizations, the International Council for Science (ICSU) and the World Meteorological Organization (WMO), proclaimed the International Polar Year in 2007–2008. It spanned 2 years, because it takes a lot of time to actually do a year's worth of work in the polar regions, and if you are going to do it in the North and the South, it takes 2 years. They convened a joint science committee to set the overall goals, and the committee immediately adopted the most advanced free and open data access policy.

Using the World Climate Research Program's program as a model, they established an International Program Office to organize and manage the event, including managing the data aspects. I had the challenge and privilege to lead that office. Then we took the crucial step of inviting individual international proposals for IPY projects. This was done deliberately to bring the best ideas forward. We did not build IPY from national plans, and we did not want to write the central plan. We wanted to solicit as many ideas as possible. This led to enormous success in the diversity of science and international participation, but it left data management as a severe challenge.

Each of these projects was intended to bring forward urgent and significant research; something that was more than what you would do in your normal routine. Each of these projects was to stimulate international partnerships. The average project had about 15 international partners. Some had as many as 40 partners. They were to build connections across disciplines between science and policy, across generations. They were to store and share data. Again, IPY set explicitly a free and open data access policy. This was not a passive requirement. Every project for endorsement had to check a box just like you check an agreement when you download software. We agreed to the IPY data policies; everyone opted in.

They were to take on substantial education and outreach. Each project had to show how to expand the polar community, by which we meant not only reach out from polar science disciplines to other science disciplines but also from polar science countries to other countries. It was a community building and outreach exercise.

We attracted enormous interest, and ended up with 230 IPY-endorsed projects related to earth, land, people, ocean, ice, atmosphere, space, education, and outreach. For geography, we have the Arctic, the Antarctic, and projects and processes that occur in both the Arctic and the Antarctic. Of the 230 projects endorsed, 170 of them were funded, and most of these projects got most of what they requested.

---

[3] Retired.

Another aspect I would like to highlight is the participation of indigenous partners, both as leaders of projects and as partners in the research team. We wanted to build the capacity of leadership from this community.

We generated $1.2 billion in science funds. That number is growing as countries start to reassess what they actually spent on IPY, and I think that number will easily surpass $1.5 billion. One of our goals was to stimulate new money for polar research, and we succeeded. There were 63 countries and easily 50,000 participants. That number we think today is probably more like 60,000. By the standard bandwidth metrics—money, people, and countries—we did extremely well in IPY.

What I want to do now is to take a more critical look at IPY. I want to assess whether we are going to meet the IPY goals and how; meaning, how are we doing today and in the future (let us say over a period of 10 years)?

- **Advance polar knowledge.** There is no question now that we had a huge impact and there will be an avalanche of polar science coming forward.
- **Enhance facilities and infrastructure**. Polar science involves huge infrastructure, such as ice breakers and bases in Antarctica. However, they are very expensive. They take a lot of fuel, and even though IPY enhanced them a lot, I think it is going to be a challenge for the polar community to keep going at the level we did in IPY.
- **Inspire the next generations.** We absolutely did that. It is hard to see how we could have done better. We have an association of 2,600 young scientists in 40 countries. They have a secretariat and a Web site. They are a recognized member of the Scientific Committee for Arctic Research and the International Arctic Science Committee. They are receiving grants from the ICSU, and we are very happy about that. They have 5 years of funding for their secretariat, but I am worried about their funding in the long term.
- **Attract public interest.** Given the resources we had, we did extremely well. In the future, I think public interest will decrease, partly because people's attention turns to other issues, but also partly because the networks that we put in place have already started to deteriorate.
- **Integrated and accessible data**. I am less optimistic here and I am scoring data low. It is that low score for data that causes me to worry about our ability to keep doing polar science in the future. Let me explain the problem. I am going to give you a tour of IPY data. It includes:
    - GPS positions, vertical and horizontal, all around Greenland and Antarctica.
    - Everything about sea ice: temperature, salinity, porosity, rheology, roughness, and extent. It includes an extensive measurement of contaminants, how they are transported, what the deposition processes are onto the ice, and all kinds of lake sediment cores, and from that, pollen records, volcanic records, and chemical records are developed.
    - For much of the work in the deep oceans, tracers, including isotopes to understand the deep ocean circulation systems.
    - Everything about ice cores: age, hardness, isotopes, dust levels, and other kinds of data.
    - On-the-ground mineralogy of exposed rock in the Arctic and especially the Antarctic, and also the microbiology of these systems.
    - Information about the social cultures of the inhabitants in the Arctic region.
    - Satellite images of new pipelines and new roads that are barriers to migration.
    - All kinds of data around the herding and meat-producing industry of the North.
    - Arctic Ocean temperature, salinity, currents, warm water under the ice, and all the aspects of that complicated system.
    - The great fish migrations of the North and the health of the fish. In this case, if you measure their ear bones and measure the isotopes, you can tell when they went from fresh water to sea water. Chemistry, physiology, and zoology are included.

- For birds: long migrations, fitness, health of the birds, and population level.
- Measurements of the sun. What are the solar properties through the dry polar atmosphere? Yesterday we talked about the atmospheric window. Antarctica is one of those atmospheric windows.

We also want to know about all the children around the world who did the IPY-wide experiments on polar days. We want to know all the students who were exposed to IPY at exhibits around the world. We want to know the number of cities and museums that had IPY events. We want to know about students who went to summer schools and gave their first poster or gave their first presentation, and we want to know the number of young scientists who have joined this new association of polar scientists.

This is just a glimpse of the IPY datasets, but you see the variety and complexity. It is literally true that, among the ICSU unions, no union could identify a science that was not included in IPY. This is its strength, but this is also its challenge from a data management point of view. Think of our data management systems on two axes: one international and the other interdisciplinary. The WMO and the World Data Centers are examples of international data activities. The WMO has meteorological and hydrological data, but it is actually relatively narrow in an interdisciplinary perspective. The World Data Centers are even more so. Canada built a relatively good interdisciplinary data portal for IPY. It does not cover satellite data or public health data, but it covers much of the range of what we would call earth science data. Canada did a very good job with interdisciplinary access.

IPY is thus both highly interdisciplinary and highly international. IPY is leading in this regard and I think this is where science is going to be in the future. It is going to be widely interdisciplinary and widely international, but we also expose the gaps. There really are no existing services and that is an institutional or international infrastructural gap.

We have projects that have good data plans and others that have adequate data plans. I define a good data plan as those projects that have a plan for both storing and sharing the data. Thirty-five of the projects have good data plans. Another 30 of them have adequate data plans. That means they know where they are going to store their data, but they do not know how they are going to make it accessible. It might be stored regionally or locally. This means that of the 170 projects, about 105 of them do not have adequate plans at all.

I lead the project office. We set the overall goals, but we do not enforce the data plans. We defer to the national organization. If you were funded by the National Science Foundation, the National Research Council of Canada, or the National Environment Research Council of the United Kingdom, that is where the enforcement happens in good data practices. If those countries have varying practices, that is, one allows a 2-year proprietary period, and one has a more aggressive and enlightened practice, then you will get that variation propagated into the projects itself. Institutional behavior inhibits the actual practice of good data stewardship. The enforcement of good data policies actually comes down to a national funding agency issue. In the IPY, where we have all these national funding agencies, the enforcement is very spotty.

We have identified that there are national issues related to enforcement of national policies, but there are also individual behavioral issues. The individual issues are related to incentives for data sharing. Elaine Collier identified very nicely a variety of reasons that people have for not sharing, but fundamentally there is no equivalent incentive for sharing. If you are competing, if you are trying to protect proprietary data, there is no incentive that actually pulls you the other way.

I wrote an article in *Nature* in early 2011, where I talked about these lessons in data sharing. I suggested publishing the data, because by doing that you can get credit for it by way of a citation, and on your promotion and tenure-record, it shows the dataset. It is very appropriate because many of these datasets involve a huge creative effort to pull together data into quality products. This can be the same effort that we put into writing a paper.

There is a new journal called *Earth System Science Data*. The journal itself is open access, and all the datasets that are published are at an open-access repository. We see that as a step forward on the data publication side.

We still have the issue of how to encourage everyone to share their individual datasets. We started to build a Polar Information Commons (PIC). This is a joint effort with the Committee on Data for Science and Technology and IPY to set up a sharing system for datasets. What we are trying to do is to set up an exchange system and to overcome barriers related to people saying, "I am afraid to share. Someone is going to steal my data. How will I know who used it?" and set up an exchange system. Below are the PIC guidelines for data users:

- You agree to cite it through acknowledgment or even coauthorship, whatever the appropriate mechanism is.
- You agree to acknowledge that you got it off of the PIC.
- You notify others of the use and of any issues that you noticed with the dataset.
- You recognize that you as the user are responsible for determining the quality and the appropriateness.
- If you make any improvements, then you return that value-added dataset to the PIC.

Notice that there is a parallel set of responsibilities for contributors. Contributors agree to make their data openly accessible. We use a Creative Commons license for that. You also agree to use the PIC badge and provide adequate metadata. You agree under limits to respond to any inquiries about the dataset, and you agree to include into the PIC system a notification of any changes that you made. The key here is not these guidelines, because they are really just standard rules of proper behavior that we already use in science. The key here is that a user can be a contributor and a contributor can be a user. As long as you are on both sides of this exchange, then there is incentive to cooperate with the guidelines and norms.

In conclusion, the strength of the IPY is that it has a whole collection of data, which is what it takes to do science in the polar region and, I would argue, in the tropical regions and the temperate regions today. It takes this kind of breadth to do modern science, but when we do this, we are going to have to bring the international, national, and individual data behaviors up to a new standard as well.

We like to say IPY polar science has had global impact. There is no question that in the short term we have had global impact on the quality of science. We had a positive effect on the public, recruited young scientists, and produced a burst of data.

23. Political and Economic Barriers to Data Sharing: The African Perspective

Tilahun Yilma
University of California, Davis, United States

Development will bring food security only if it is people-centered, if it is environmentally sound, if it is participatory, and if it builds local and national capacity for self-reliance. These are the basic characteristics of sustainable human development. James Gustave Speth (UN Development Program, 1994).

I think you have heard a lot about the problems we are facing in Africa with data sharing. I would like to share with you information about the work of Africans who have come to the University of California (UC), Davis. They have generated very significant scientific data. So if African scientists can do this excellent work at UC Davis, why are they not capable of doing the same in Africa?

The answer may be because of political problems. For example, I was not able to go back to Ethiopia, so I established an International Laboratory of Molecular Biology (ILMB) at UC Davis in 1991. The goals of the ILMB are to (1) promote a culture of science in developing countries; (2) conduct research in tropical viral diseases, including the development of recombinant vaccines and rapid diagnostic kits; (3) transfer technologies to developing countries; and (4) address the issue of barriers to the growth of science in developing countries.

Any type of aid program that does not lead eventually to self-sufficiency is actually destructive, just like welfare, and that is what has happened with many aid programs in Africa. Much of what we have heard about Africa so far is negative. Yesterday, Dr. Yang brilliantly showed how China has advanced very far both technologically and economically in the past 30 years. My goal has been to promote similar development in Africa. My idea was for African scientists to be trained at UC Davis and then to go back to train other Africans, thus allowing the successful transfer of technology and enhancement science. Unfortunately, there were many political problems and barriers that interfered with this goal, and I would like to discuss these with you.

I am going to use rinderpest as an example. Rinderpest is an important disease that played a very significant role in the development of the veterinary profession. It was the virus that was introduced into East Africa through collaborations between the British and the Italians in the 1800s. It was used as germ warfare, and in fact, 32 to 60 percent of the population and more than 90 percent of all ruminant animals, including cattle, goats, and sheep, perished.

I was sent to the United States to become a veterinarian and then returned to Ethiopia to aid in the attempt to eradicate rinderpest using a tissue culture vaccine. In the 1970s we managed to vaccinate more than 120 million cattle, and we then celebrated the eradication of this disease. Unfortunately, rinderpest was not eradicated in African wildlife; it did not take very long for the virus to spread back into livestock and once again become endemic in Africa.

The major problems associated with the failure of the first attempt to eradicate rinderpest from Africa included the lack of heat-stability, the expensive cost of producing the vaccine, the difficulty in administering the vaccine, and a lack of continued surveillance for the disease. I decided to develop a safe, heat stable, inexpensive, and effective vaccine by implementing the new technology that uses the smallpox vaccine (vaccinia virus) as a vector for recombinant vaccines to express the protective proteins of the rinderpest virus. Using vaccinia virus as a vector, we developed several vaccines for rinderpest; the first paper demonstrating the safety and efficacy of this vaccine was published in the journal *Science* in 1988. Further refinements of the vaccine led to field trials in Africa and demonstrated that it can be given intramuscularly or intradermally. We also developed and transferred an inexpensive, companion diagnostic kit that can be used to distinguish vaccinated from infected animals. This is what African scientists have done in collaboration with scientists from the United States and developing countries at the ILMB.

Other vaccines we have developed utilizing this technology include a vaccine for a disease similar to foot-and-mouth disease, vesicular stomatitis. Unlike the rinderpest vaccine, the level of protection was very minimal. Another vaccine we developed in our lab was for HIV; we were the first to show that vaccines for HIV do not protect but only reduce the virus load in the blood. This was one of the three major contributions considered in the HIV field. These results indicate that many recombinant vaccines require enhancement of efficacy to be protective immunogens. What we needed then was to enhance the efficacy of these recombinant vaccines.

During his inaugural speech, David Baltimore, the Nobel Laureate and former president of the California Institute of Technology, stated, "When you grow up with a world like that, there is a central aspect of society that makes no sense: politics. For years, I simply could not comprehend what that meant. When people said that in making decisions, you need to consider both the rational elements of an issue and the political ones, I did not understand what they meant—why was not rationality enough? So my whole life since I left my parents' nest has been an education in irrationality. I have had to learn that you cannot deny the passions of people, you must accommodate them; that you cannot deny history, you must accommodate it. I think this is a perspective that all scientists who are willing to work within the larger society have to learn, and it is what sometimes limits the effectiveness of scientists when they do venture outside of their laboratories and institutions."

This is really what I have learned. Using science to make vaccines is simple, but dealing with the politics is very difficult.

The *Journal of Virology* is the number one international specialty journal for virology. Details about the rinderpest vaccine that was tested in Kenya were published there by a group of scientists from the International Laboratory of Molecular Biology for Tropical Disease Agents who originally came from Argentina, Brazil, Afghanistan, Pakistan, the United States, Ethiopia, and Kenya. You can see that these people who came together were able to develop what Dr. Gordon Ada described in the journal *Nature* (January 31, 1991)[4] as one of the two outstanding recombinant vaccinia virus vaccines in the world. These results help prove my point that people from developing countries, if given the right opportunity, are quite capable of competing with scientists in developed countries.

Another contribution that people from developing countries have made is to use cytokine genes for enhancing the safety of vaccines by more than 100-million-fold. One brilliant Ph.D. student born in Afghanistan, published her dissertation research on the topic in the *Journal of Virology* and the *Proceedings of the National Academy of Sciences*. Another Ph.D. student, from Ethiopia, has advanced a concept to develop a safer, effective vaccine for smallpox. Extending this work, we have developed a recombinant vaccine for Rift Valley fever, a disease that affects both humans and livestock, a project sponsored by the U.S. Department of Homeland Security. Rift Valley fever virus is considered a very dangerous agent that could be used as a bioterrorist weapon. We have also developed diagnostic tests for both Rift Valley fever and foot-and-mouth disease.

The program that we started at UC Davis is to advance self-reliance. Based on that, we built labs in Egypt, Kenya, Ethiopia, and Senegal. An insect virus expression system was used to produce a recombinant rinderpest protein used in the development of a diagnostic kit at the ILMB. This reagent is

---

[4] Available at http://www.nature.com/nature/journal/v349/n6308/pdf/349369a0.pdf

produced at high levels and does not come from a virus capable of infecting mammals. Thus, what would have cost $60,000 could be produced for 5 cents. African scientists trained at the ILMB went back to Senegal, produced the diagnostic kits, and trained other African scientists from 30 different laboratories to successfully transfer the technology.[5]

The accomplishment of postdoctoral researchers and students in the lab is evident in publications such as *Science*, *Nature,* and other top scientific journals, including *Nature Biotechnology*. This is proof that these people are capable of making valuable contributions in a supportive environment. We have received the highest award in animal science, been elected to the National Academy of Sciences, and honored with the University of California Medal. All this was done by people from developing countries.

Then the question I ask again is, why not in Africa? If we can do this in Davis, California, why can we not do it in Africa? I am not asking about Southeast Asia, because progress there is obvious. One of these days, I hope we can say the same thing about Africa.

What are the barriers internally and externally that prevent the development of science in Africa? Natural resources are sucked dry by governments from developed countries. Moreover, African countries pay billions for military goods and warfare. When I worked in Senegal, I asked a colleague, how is it that you have absolutely no natural resources, yet you have a very high per capita income in the African continent? Her response was, "We are blessed in Senegal in that we have no natural resources. Thus, we are left alone and spared from destruction."

One example I like to use is the war that was conducted between Eritrea and Ethiopia. People from both countries speak the Tigrinya language. Each country spent more than $2 billon purchasing military weapons and jets to fight a war for a piece of desert called Badme and sacrificing more than 200,000 people. According to *Africa Today,* observers likened the conflict to "two bald men fighting over a comb." What economic or strategic benefit could be gained from control of 400 square-kilometers of a rocky triangle of land over which these two former allies were now locked in battle? "Eritrea already has enough rocks," says one analyst, adding that "if rocks were worth money, Eritrea would be the richest country in the world."

In my opinion, the countries in Africa should follow the example of China, India, and Brazil if they want to achieve development and overcome their reliance on destructive foreign aid. Africans should work toward becoming more self-sufficient.

---

[5] Yilma, et al. (2003) Inexpensive vaccines and rapid diagnostic kits tailor-made for the global eradication of rinderpest. In Vaccines for OIE List A and Emerging Animal Diseases. Developments in Biologicals. 114: 99-111.

PART FOUR: LIMITS AND BARRIERS TO DATA SHARING

24. DISCUSSION BY THE WORKSHOP PARTICIPANTS

PARTICIPANT: I am going to ask a question as a former president of the Committee on Data for Science and Technology (CODATA), which is an international organization concerned with advancing the use of data within science. We have heard a number of talks about various barriers, and one of the questions that we have been grappling with for a long time is what we can effectively do to break down some of these barriers, given that in the United States you buy a personal computer and you basically are ready to be a data scientist. In many parts of the developing world, that is not the case. They do not have the connectivity, software, or training to do that. What practical steps could an organization like CODATA take to actually help break down some of the barriers that we have heard discussed today?

DR. RILEY: If your university was spending 30 percent of its budget on Internet connectivity, how would you feel as a faculty member? You would probably grumble. You would probably go to your faculty governance meetings and say, "We have to do something about this." One of the interesting things is that as these National Research and Education Networks evolve, and as this international connectivity comes into place, and networks are built and the prices start to fall, you start to free up money that can then be used on other things, like computers and software and content-related activities. It is just amazing when you look at how much money is being sucked out of the universities because of this connectivity problem. We have recognized that something needs to be done. One of the reasons I mentioned the statements from the InterAcademy Panel is that we spent a long time trying to craft messages asking, "What can you as members of the Academies do? We would like you to refer your governments to these statements. Go into your societies, and ask for help in fixing this problem so we can get on to the things we really want to do."

DR. CARLSON: We cannot accept the status quo, and CODATA can be an agent to identify that interface. A good idea like an information commons rapidly runs into pushback from the research infrastructure: legal pushback, behavioral pushback, and so on. CODATA can be the agent to keep pushing forward.

DR. COLLIER: I will comment from the perspective of the National Institutes of Health (NIH). The National Center for Research Resources supports a lot of development in rural parts of America. There were and still are real connectivity problems in parts of this country. I am amazed by the creativity of some of the people at those institutions who solve problems by using lower-technology solutions that actually move data. I think that we need to push forward on getting the higher performance and bandwidth, but we also need to not squash creativity and actually allow things to happen with what is available. Just saying you cannot do it until you get more is not true in many cases.

DR. RILEY: I need to clarify what I was saying. I was not saying that nothing is happening. There is a lot of great activity going on, but what could happen if we could free scientists in less-developed countries from the low network connectivity they have? It is amazing how creative people everywhere have been trying to get access. However, try doing your research onsite in the middle of Botswana, for example, and you will understand what the limitations are. What could these great students and faculty do if they had the kind of connectivity I have at home, which I do not think is even as good as when I am sitting in my office, but it is wonderful compared to what they have.

DR. KAHN: My question is directed to Professor Riley. On one extreme, he is talking about a poverty of technology, and he has just reemphasized that, whereas at the other extreme, Dr. Yilma is talking about the poverty of politics. Dr. Yilma said that in Africa there are not enough people to manage the routers and networks. Yet Africans have been more than capable of putting in the routers and the networks and running mobile telephony and making fortunes. Here are some data. The 300th wealthiest person in the world is an Egyptian named Naguib Sawiris. He runs a network. Another example is the 630th in the world, who is a former Sudanese and now lives in the United Kingdom. His name is Mohamed Ibrahim. It seems where there is a demand, the technology is not the barrier. I would like your comment in reply.

DR. RILEY: Cellular technology is generating a revenue stream that is now making it economically feasible to use submarine cables, and a small number of people are getting wealthy. They hire a few people to do those routers and switchers, and a lot of them are nonnative Africans. Many countries in Africa are using Chinese money, including loans to bring in Chinese engineers to build national backbones, and also highways and mines and petroleum centers using Chinese technology. When it is over they will be captive to owing the Chinese government money, and captive to the annual maintenance costs associated with the equipment. They will not have had much development of the human side of being able to run the routers and switchers other than maybe what ties into those cell towers. The problem is very deep and very complex. I would stand behind everything I said, because most of it comes from colleagues in Africa or people doing that infrastructure. There are new forms of colonialism going on, too.

PARTICIPANT: There were two messages that I think were not consistent in Dr. Yilma's speech. One message was that Africa is better off being left alone. The other message was that places like the University of California, Davis, and Columbia University have resources and training and capacity to help development constructively without leaving Africa alone. An example is my center, where we do a lot of capacity-building training on how to manage data and use data for decision making in Millennium Villages and in the African context. I am just curious which message you really think is the one that is important.

DR. YILMA: I would be happy to address this. Africans lived for a millennium in harmony with the environment. Since the colonial introduction, look at what has happened to Africa. It is in total turmoil.

PARTICIPANT: I was involved in the selection of the Millennium Villages from the ecological and sustainable development point of view. As far as I know, the villages do not have direct links to the people involved in the conflict. The approach of the whole network of Millennium Villages is to help build capacity and work with expertise in the developed countries. We have a big project in Haiti, which is a postconflict country just like ones that are currently in conflict. The point is not to ignore the concerns of postconflict countries. Those are the countries that in fact need help and need positive reinforcement separate from these other political forces. You did not really answer the question of whether you believe in groups at U.S. universities and other developed countries getting engaged with the developing world. Is that not a model that you also think is one that can work?

DR. YILMA: Absolutely. I support such activities, and I think there should be an exchange of data and exchange of scientists. This is a wonderful thing.

PARTICIPANT: We have this improvement taking place in connectivity and hopefully it will also change the possibilities for Africa. Simultaneously, however, we have a growth of databases and we get very big amounts of data. It is really hard to work with these databases if you do not have access to very good broadband. My question is: Do you think that we should consider the problems for lack of connectivity in certain parts of the world when working on the design, structure, and accessibility of these databases so they can be easily accessible? Is that a problem?

DR. RILEY: The answer is, of course. One of the problems has been that people who do collaborative projects with African colleagues often make a decision to put their data on a server back home. The consequence of that is that their African colleagues then have a hard time getting the data. As this connectivity comes into place, this is going to create new opportunities where that does not have to be the case. For example, initially the submarine cables will land at places like Mombasa, Dar es Salaam, and, of course, South Africa. The University of Dar es Salaam has been one of the first to get some of that bandwidth connectivity, and I think Kenyatta University and several in the Nairobi area have as well. Those become interesting places to think about hosting data centers as the infrastructure gets built up around the rest of the country.

PARTICIPANT: We heard in some of the presentations yesterday and today that at least one of the barriers to sharing data are the lack of incentives and motivation from scientists. Professor Carlson, you articulated very nicely some suggestions for changing professional institutional structures to get scientists to share data, but in my field in global health and public health, the incentives are not enough. You cannot just have carrots. You have to have sticks. For example, the NIH, Canada, a lot of public funders, and even private funders in my home country require submission of data-sharing plans, but there is not that follow-up or at least a robust system of enforcement. I was wondering what the panel thinks about having something like sticks and what those sticks might look like. If incentives are not enough to motivate data sharing, what should the penalties look like? Is that a good idea?

DR. CARLSON: I completely agree. If you look at the regulatory environment of the funding agencies and talk to the directors of those funding agencies, most of them would say, "Yes, we have those sticks in place. If scientists want to come back to us for a renewal of their grant, it already says in the grant award they must have a data plan and provide their data to a center." The problem is not the lack of the regulations. The problem is that at the program manager level, there is no enforcement. I can say this honestly without listing any country. I have been into a research council where the director said, "We have in place procedures so that our investigators have to provide their data in order to get a renewal. However, the program managers have no idea if any of their principal investigators have provided that data. They are not set up to track it. They have no mechanism, and there is no incentive to track it." It is actually a culture within the agencies. The agencies already have the written regulations, but the agency culture has to change.

DR. COLLIER: I think that you can do some things with sticks, but if you did the enforcement, where would these people put their information? How would it be sustained? Do we have the infrastructure in any country to deal with that? Who do we hold accountable? Is it the institution? Is it the investigator? What if the investigator dies? What if the institution goes bankrupt? We do not have a real plan for how we are going to do this. I think that you can push, but it is going to take pushing in other places and providing some solutions as to how we are actually going to manage that information, where it is going to be, and how we are actually going to access it.

PARTICIPANT: I am wondering if the panel would be willing to address two issues that are enormously hindering, not only to data sharing but also to general science development in emerging economies. One of them is taking again what Michael was saying about cellular telephony as well as broadband availability. He mentioned a number of individuals, and we also know that the first or second richest man in the world is a Mexican who controls telecommunications throughout the continent. What we have experienced is that telecommunications in all of South America are enormously more expensive than they are for comparable services in the United States. That is one of the issues.

The other one is somewhat related, although it is strictly applied to the biological sciences, and that is the issue of intellectual property rights (IPR), either for therapeutic agents or for vaccine agents. There are a variety of agents that can be produced through recombinant technologies at extraordinarily low cost compared to the price at which they are available commercially in Africa and Southeast Asia. There is quite a bit of controversy regarding the prudence or the rationale for proceeding with producing the necessary agents regardless of IPR. Two common issues link these two: one is enormous greed, and the second is corruption. Would the panel be willing to address that?

DR. RILEY: How do you deal with greed and corruption? I am not sure that we know how to do that. Look at what is going on around the world right now because of people getting fed up with greed and corruption in their countries, and now using these tools that seem to be unstoppable, although there certainly have been attempts to stop them by shutting down cell phones and shutting down the Internet. If we could get rid of the devil, we would solve the problem.

PART FIVE:
HOW TO IMPROVE DATA ACCESS AND USE

25. Government Science Policy Makers' and Research Funders' Challenges to International Data Sharing: The Role of UNESCO

Gretchen Kalonji
UNESCO, France

I am going to offer you a very brief overview of the United Nations Educational, Scientific and Cultural Organization (UNESCO). I will give some examples of current activities that have to do with the challenges of data access and sharing, particularly in the developing world, and then proceed with some ideas about new opportunities and new directions that we might pursue in hopes of getting feedback from you and even perhaps forging new collaborations.

UNESCO was founded in 1949, and has an extraordinarily broad mandate covering education, science, and culture; communications and information; and ethics and philosophy. Such a broad mandate could be seen as a disadvantage, but within the context of the challenges that we are addressing at this meeting today, I hope to show you why this mandate may in fact be a very useful thing.

The organization has some strong existing programs within the natural sciences sector, in particular, the well-known Intergovernmental Oceanographic Commission and the International Hydrological Program. Both have been around for 50 or so years. We also have the ecological sciences with the Man and the Biosphere Program and the International Geosciences Program. What is perhaps less well known is our strong focus on indigenous knowledge and science policy. Science policy is one of our largest areas.

We are headquartered in Paris, but have science offices in Jakarta, Nairobi, Montevideo, Venice, and Cairo. We have science officers in about 53 UNESCO offices around the world. It is a strong, geographically distributed network with people on the ground actually working on projects.

UNESCO has a number of affiliated institutions, including our Category One Centers. The best known to this community is probably the International Center for Theoretical Physics in Trieste, but we also have an international hydrological education program in Delft, which is the world's largest postgraduate freshwater program, with 80 percent of the students coming from the developing world. The Academy of Sciences for the Developing World, TWAS, is also a UNESCO institution.

UNESCO also has what are called Category Two Centers, which are affiliated research centers established by our member states within a particular country and funded by that country. The country agrees that they will take on an international responsibility for a particular topic (e.g., water-based disasters, including one in Japan called ICHARM [International Center for Water Hazard]). We have about 22 of those in the sciences. Most are in water, and there are four new ones being established in Africa. Lastly, we have UNESCO Chairs around the world, which are appointed in a competitive process, and they are another wonderful resource for UNESCO.

One of the things that we have that is particularly important for the challenges we are discussing today are the extraordinary and very well-known World Heritage sites. They are designated for either cultural value, natural value, or both. In addition to the World Heritage sites, we have the biosphere reserves and the newly emerging geoparks, which are very popular in some countries, particularly China. Those are areas where a combination of research and education and community economic development can take place in an integrated manner.

We also have a network of affiliated partners. CERN, the European Organization for Nuclear Research,[1]

---

[1] See http://public.web.cern.ch/public/.

was in fact created through UNESCO, and we continue to work very closely with them on issues such as digital access in Africa and physics education for teachers in Africa. The International Council for Science (ICSU) predates us but is a very close partner institution. The International Union of Pure and Applied Chemistry is another partner. We are working very closely with them on the International Year of Chemistry. One of the things that is perhaps less well understood about UNESCO is the very close relationships we have with our member states. We are unique within the United Nations specialized agencies and programs in that we actually work in the same building in Paris as the permanent delegations from the member states, which enables a very close working relationship on concrete projects. We also have a structure that is unique within the UN system. The national commissions for UNESCO bring civil society together to help set directions for the organization. These commissions are more or less active. Korea, for example, has 600 people working on education, science, and culture, and UNESCO has become a household word there.

Lastly, we have perceived political neutrality. What that means is that we can convene discussions about topics that are quite thorny and have our 193 member states from around the world come together and discuss them in an amicable manner.

On the other hand, we could have a more strategic focus. We need to have a better working system of all of these various parts and partners; we need to work better with other UN agencies, the private sector, and other sectors of society; and we need to enhance our visibility.

Given these strengths and weaknesses, UNESCO's science agenda should prioritize three things. First, we should help tackle problems that intrinsically require international cooperation and provide services for member states in that regard. Second, we should build on our original mandate. UNESCO was created with the slogan of building peace in the minds of men and women. We focus on those areas in which the science agenda interacts very closely with the issue of conflict prevention and conflict resolution. The broader agenda of peace is very dear to our hearts. I cite a couple of examples here. One is our work on transboundary aquifers, which I am going to talk about later in terms of large-scale data challenges. The other one is a fascinating effort called SESAME (International Center for Synchrotron-light for Experimental Science and Applications in the Middle East), which brings together scientists throughout the Middle East. It is a very important project in that it is putting together a synchrotron in Jordan and bringing together a scientific community from throughout the disciplines that can use this light source. It is an extraordinary example of scientists from a region with a huge amount of tension actually working together. Iranians, Israelis, Palestinians, Turkish, and so on are all working on the same large-scale science project.

Lastly, perhaps the bulk of our activity falls into serving the member states. We are international civil servants. We need to do the best job possible to help our member states reach their own goals for building scientific, technological, and innovation capacity in order to address poverty eradication and also provide the scientific basis for the solutions that are being proposed. Of course, we should continue to prioritize work in those areas where we have a lot of expertise, such as water.

At UNESCO, we have a unique view regarding the science and the development agendas. We have a very people-centered approach—an approach that is based on empowerment, ethics, and respect for local knowledge, but also our conviction that the ability to contribute to global challenges and the opportunities to do so are in fact fundamental human rights.

Since I joined UNESCO, we have melded our activities into a new strategic plan to be approved by our executive board next month. We have clustered our activities into two main areas, which I will talk about to show how the data challenges map onto some of these activities.

One area is strengthening science, technology, and innovation ecosystems. Societies have to have good policy, and we put an enormous amount of attention into that effort, particularly in Africa.
Some of the symposium speakers have stressed that universities are really the heart of healthy technological innovation ecosystems. We have a big focus on higher education in the developing world, and on mobilizing the popular understanding and support for science, such as science for parliamentarians, science journalism, and working with science museums, as well as programs that enhance participation of indigenous people. To summarize, this cluster focuses on

- Promoting science, technology, and innovation (STI) policies and access to knowledge;
- Building capacities in basic sciences and engineering, including through strengthening higher education systems; and
- Mobilizing broad-based societal participation in STI.

The second cluster is mobilizing science for sustainability. This is an activity where large-scale scientific communities come together to set a collective scientific agenda.

I want to discuss a couple of examples for how these large-scale data issues become the actual work of our UNESCO family. In freshwater, UNESCO has the International Hydrological Program, which is an intergovernmental effort. Each nation has its own committee that works on setting a collective agenda in the area of freshwater. Then together they develop a 6-year plan that they modify over time. This is just one of the examples in which a community of hydrologists working together is trying to assemble the kind of data that we need. An example of the success of their work is in the transboundary aquifer in parts of the world, including Africa.

In this kind of an effort, UNESCO plays a coordinating and somewhat catalytic role, but basically there are multiples of hundreds of hydrologists around the world working on a common agenda. This is very important for avoiding conflict. Our connections with the UN system means that the scientists can work with the legal people in the United Nations and the diplomatic representatives to help forge the law in the general assembly concerning the equitable sharing of transboundary aquifers.

In the current work plan for freshwater intergovernmental science programs, there is a big emphasis on education, sustainability, basic sciences, and climate change. There are also cross-cutting programs, such as networks of hydrologists who work on a regional and global basis sharing data for hydrological research. For example, there is a Nile River basin group that brings together the scientists who are dealing with the Nile River water issues.

Another example of data sharing that is qualitatively different is the Man and the Biosphere Program. In this program, there are 564 sites in 109 countries. These sites are proposed by each country. There is an intergovernmental body that decides whether it can become a biosphere reserve. The interesting thing about the biosphere reserves is that, unlike the World Heritage sites, they involve a region that is protected because of biological diversity, but humans also live there. There is also a buffer zone surrounding the core region, and an extended zone. What that means is that activities such as mining, tourism, and farming are not forbidden. It gives scientists the opportunity to have some very vibrant case studies of the international balance between biodiversity conservation and economic development and livelihoods for local communities.

Let me briefly touch upon some other areas. UNESCO's science policy activities range from international, like the *World Science Report* and the *World Engineering Report*, to regional and country-based policy support. Twenty-two countries in Africa are working with us right now on their science policies.

In the geological sciences, there are many examples in which the global change research community has been working together with a broader geological community, particularly the International Union of Geological Sciences (IUGS), on getting access to and combining and sharing geological information from a variety of sources, including paper-based sources. It is a very exciting time now for the biodiversity community, too, and an intergovernmental platform on biodiversity and ecosystem services has been discussed. The biodiversity analog of the Intergovernmental Panel on Climate Change (IPCC) will undertake a very large-scale effort to promote access to data in the area of biodiversity. This is particularly exciting because of the newly created Nagoya Protocol for Access and Benefit Sharing.

There are three qualitative areas to which I believe UNESCO contributes. First, it helps strengthen the capacity of member states to engage in data-intensive science. Second, it provides platforms for more effective community engagement. Lastly, UNESCO enhances awareness of the value of freely sharing scientific data.

UNESCO could, if it is of interest to other partners to work with us, potentially host a meeting in Paris with our member states about the same topic, because they are the direct representatives to the government. They are the ones who need to hear the speeches like the one from Professor Yang about how wonderful it was for China to make data freely accessible.

My second idea is to incorporate within our existing efforts on strengthening higher education a collaboration on developing capacity in data-intensive science in partner universities, especially in Africa. It should be very straightforward to integrate awareness-raising activities into some of our existing efforts, like our work with ICSU in preparation for the UN Conference on Sustainable Development (Rio+20), or programs on science for parliamentarians, or our work on policy.

Lastly, I am very excited about the Intergovernmental Science Policy Platform on Biodiversity and Ecosystem Services (IPBES). It seems very likely that UNESCO, together with the UN Environment Programme and maybe another agency, will be taking the lead as the institutional cohost for IPBES. I would be interested in brainstorming with individuals or organizations about this extraordinary opportunity.

26. International Scientific Organizations: Views and Examples

Bengt Gustafsson
Uppsala University, Sweden
ICSU Committee on Freedom and Responsibility in the Conduct of Science

I will begin with some historical remarks. History provides an enormous data bank, which is also useful when we discuss the accessibility of data banks in contemporary research. By starting the discussion by referring to the development in Europe some four to five hundred years ago, we find scientists quite often keeping their truths between themselves, and even sending cryptographic messages to each other to prevent others from reading or understanding. The interesting counter-examples in the early seventeenth century were the new artisans, the small factories, and the people developing technology for mining. They were open-minded, and symbolized modernity.

Openness was from the very beginning connected to the idea of progress—progress in arts and in building a new society. That was clear and strongly brought forward by Francis Bacon. Let me cite from an account of this by William Eamon in the *Minerva* article, *From the Secrets of Nature to Public Knowledge: The Origins of the Concept of Openness in Science*: "One of the lasting effects of the influence of Bacon's philosophy was the establishment of a new model of the scientific research worker as one dedicated to the pursuit of knowledge for the public good. No longer was science to exist merely for the pleasure and illumination of a few minds; it was to be used for the advancement of commonwealth in general. This new demand required that more knowledge be shared, both within the scientific community and with society at large."[2]

However, this was not the beginning; traditions along these lines existed before the Renaissance in Europe. But the ideas from Bacon's time form the ideological tradition in which we scientists still live and work. This was also the spirit in which the Royal Society was formed in 1660, in fact, directly inspired by Bacon and his writing, and that also pursued the idea of transmitting publicly the findings by its Philosophical Transactions. There were many academies formed on this model. One was my own academy, the Royal Swedish Academy of Sciences, in the following century. The aim was to develop and spread knowledge in mathematics, natural sciences, economy, trade, useful arts, and manufacturing. There was also the idea of publishing descriptions of research achievements and creating a pregnant almanac containing advice for farmers and others, which was the second book printed in Swedish during more than 100 years. Only the Bible was read more than the Academy almanac.

These ideas are important cornerstones in the foundation of all academies still and also for the International Council for Science (ICSU), which has about 100 national members, most of them academies, as well as some international scientific unions. Since the 1930s, ICSU has built its activity on the Principle of the Universality of Science (Universality Principle). This principle is fundamental to scientific progress. According to the 5th statute of ICSU, the principle involves freedom of movement, association, expression, and communication with scientists, as well as equitable access to data, information, and research materials. In pursuing its objectives for the rights and responsibilities of scientists, ICSU actively upholds this principle and, in so doing, opposes any discrimination on the basis of such factors as ethnic origin, religion, citizenship, language, political stance, gender, sexual orientation, or age. ICSU states that it shall not accept disruption of its own activities by statements or reactions that intentionally or otherwise prevent the application of this principle.

What can we do to uphold this principle in reality? ICSU has taken a number of steps for this, such as

---

[2] Eamon, W (1985). From the Secrets of Nature to Public Knowledge: The Origins of the Concept of Openness in Science. *Minerva* 23, pp. 321-347.

establishing committees, including its current Committee on Freedom and Responsibility in the Conduct of Science (CFRS). It is clear from the name of this committee that it has wide objectives, reflecting the view that the freedom advocated in the Universality Principle should also require that important responsibilities are taken by the scientific community relative to the society. In addition to this committee, ICSU has also taken other initiatives, which are directed more particularly toward securing data distribution and data accessibility, such as initiating a Strategic Coordinating Committee on Information and Data, looking at the interaction of the World Data System, the Committee on Data for Science and Technology (CODATA), and other ICSU data- and information-related activities. ICSU has also cooperated with other organizations in forming an International Network for the Availability of Scientific Publications.

The CFRS discusses and takes stands against breaches of the Universality Principle. This is often done in collaboration with several other organizations, in particular, the members of ICSU. Another important collaborator is the International Human Rights Network of Academies and Scholarly Societies. The CFRS advises ICSU and ICSU members on related matters and helps arrange conferences and workshops on issues of responsibility and integrity of science. In doing so, it is important for the committee to also try to reach conclusions; after such workshops and conferences, conclusions are published as statements or advisory notes. Some recent examples, in addition to this workshop, is a workshop on access and benefit sharing of genetic resources held in Berne in June 2011, as well as one on private sector–academia interaction in November 2011 in Sigtuna, Sweden. Other workshops are planned on science policy advice, science and antiscience, and science in contemporary wars. All these workshops must be truly international and will have a focus on the balance of responsibility and freedom in science.

Now, turning more particularly toward the question of access to scientific data, it must be stressed that the present situation, although improving, is far from satisfactory. As Paul Uhlir pointed out in a 2010 essay in the CODATA *Data Science Journal*, there are still "information gulags," that is, large numbers of data resources in dark repositories; "intellectual straitjackets," exclusive property protection of data when not needed; as well as "memory holes," meaning that data once collected are often later destroyed or not conserved properly[3]. Existing data are most often not properly archived. Even quite important data are not maintained. Early National Aeronautics and Space Administration (NASA) and National Oceanic and Atmospheric Administration (NOAA) data are examples of this.

Let me now focus on my own field, astronomy, and provide several examples to show the progress in openness, with a more historical twist than that of the stimulating presentation by Željko Ivezić. A famous example of secrecy in science is the behavior of Galileo Galilei when he summarized his pioneering telescopic studies of the planet Saturn in the early 1610s. He did not interpret what he saw as a ring system, since there seemed to be two bodies on either side of the planet, or possibly "ears" on the planet. Yet, to claim his discovery in spite of the uncertain interpretation, he used an anagram (when deciphered indicating that the planet "had triple form") as a form of commitment scheme. Not until several decades later, Christian Huygens correctly identified the ring system and announced its existence.

Twice a century, Venus passes across the solar disk; the first detailed European observations of this phenomenon were made in 1639. James Gregory next proposed that a method could be developed to find the distance to the sun by timing exactly when Venus enters the solar disk and when it leaves the disk. If that is done from several places on Earth, the distance to the sun could be determined accurately. Edmond Halley proposed that astronomers should observe the Venus transits systematically next time, which happened to be after his death in 1742. Astronomers traveled to various parts of the world to measure the transits of Venus accurately. The first passage was in 1761. There were even measurements made from a

---

[3] Uhlir, P. (2010, October 7). Information Gulags, Intellectual Straightjackets, and Memory Holes: Three Principles to Guide the Preservation of Scientific Data, CODATA. *Data Science Journal, 9.*

naval frigate while fighting pirates in the Mediterranean. The data were then assembled, shared among a network of astronomers, and made common.

During the event 8 years later, astronomers were even better spread across the globe, including a British expedition on Tahiti where young James Cook was sent with his vessel to carry out the observations. His observations turned out to be rather poor. The excuse given by Cook was that the quadrant was robbed by a local chief and dismantled, and then only provisionally mended. On his way back, however, he discovered some new regions of the earth and reported this back to the admiralty, and they forgave him explicitly for his bad observations. The data were assembled and discussed among the astronomers, because they did not have the quality expected—the distance calculated by Jérôme Lalande in Paris was 153 million kilometers plus 1 million kilometers, which was not as good as they had hoped for. This very problem of coordinating all observations and minimizing the errors led to requirements for further openness.

Another great international project from the following century was the great star map, *Carte du Ciel*, where 22 observatories joined in constructing rather similar telescopes, which together exposed 22,000 photographic glass plates with 4.6 million stars to be measured and printed. Many people, including non-expert women, were engaged in these activities. The whole result was published in 254 volumes. Again, the need to reach the goal required wide international collaboration, and to achieve optimal quality, openness was necessary. This experience that such ventures must be carried out in common, not only to make the heavy workload possible but also to achieve an optimal analysis of the data, demonstrated to the astronomy community the importance of sharing data.

In contemporary astronomy we find this intimate collaboration taking place not only in discussing the data or sharing data but also in the very setting up of projects and determining how to analyze the data, how to release them, how to publicize them, and how to provide assistance so as to allow as many people as possible to contribute. At many of the largest international telescope facilities, financed by consortia with universities or states as members, nonmember astronomers may take part, and in some cases even be principal investigators (PIs) on projects. At least in principle, only excellence of the project proposal matters. Finally, in most cases, all data are made fully public after about 1 year.

We can compare this situation with CERN, the European high-energy physics laboratory at Geneva. Nonmember state PIs work there and are playing important roles. CERN also runs an open fellowship program, to which scientists from all over the world are encouraged to come and team up with others. Some primary data are available from earlier experiments, but the recent ones, including the Large Hadron Collider experiments, will probably not be available publicly, just because they are so extensive. It is very hard to interpret them without being a member of the experimental group. In this case the enormous database of primary data may still be closed.

Astronomers produced a data manifesto 5 years ago, which was later adopted by the International Astronomical Union (IAU). It starts with this declaration: "We, the global community of astronomy, aspire to the following guidelines for managing astronomical data, believing that these guidelines would maximize the rate and cost effectiveness of science discovery."[4] Relevant guidelines include: "All significant tables and images published in journals should appear in astronomical data centers. All data obtained with publicly funded observatories should after proprietary periods be placed in public domain. In any new major astronomical construction project, the data processing, storage, migration and management requirements should be built in at an early stage in the project and budgeted along with other parts of the project. Astronomers in all countries should have the same access to astronomical data and information. Legacy astronomical data can be valuable and high-priority legacy data should be preserved

---

[4] Available at http://www.atnf.csiro.au/people/rnorris/papers/manifesto.pdf

and stored in digital form in the data centers. IAU should work with other international organizations to achieve our common goals and learn from our colleagues in other fields." Legacy astronomical data refers to older data. For instance, the glass plates mentioned above can be valuable and even high-priority legacy data, and they should therefore be preserved and stored in digital form in the data centers.

Implementing such a manifesto is, of course, a major undertaking. One important initiative along these lines is the International Virtual Observatory Alliance, which is a worldwide organization of national members, which tries to make all astronomical observational data in various wavelength bands from various instruments available publicly. Several international astronomical data centers also play a great role in these endeavors.

Such efforts to promote openness are not only for generosity. There have been studies showing that the open data policy of the Hubble Space Telescope (HST) has increased the number of publications based on HST data by a factor of 3, and that the earlier satellite telescope International Ultraviolet Explorer has increased the number of publications based on those data by a factor of 5. Another important issue on openness is open-access publications. Most astronomy papers can now be accessed freely via a preprint database or archive (Smithsonian/NASA Astrophysics Data System), and most major journals accept this way of prepublication.

Let me also comment on the problem of overcoming the digital divide. There are several concrete examples of attempts within the scientific communities in particular fields to bridge the gap between the developed and the developing world in this respect. In astronomy, this bridging has partly been driven by the important fact that many of the best astronomical sites are situated in developing countries. The IAU has recently established an office for astronomy development in the developing countries at the South African Astronomical Observatory. The experience at the International Science Programs at Uppsala University, which has been actively bridging between university science departments in the developing world and in the Nordic countries for 50 years, shows that much can be done to diminish the digital divide with patience and consistency.

No doubt, astronomy is a simple example with a long tradition of international collaboration and openness. Astronomers have realized that international collaboration is necessary, because the universe is large and rich in a multitude of various phenomena. Astronomy has limited economic impact and interest and few security restrictions. And it is mainly motivated by the interest of the public, which after all have to pay for it, and is why openness is necessary. So, astronomy is a simple case. Nevertheless, we may learn from it. There is a full chain of openness aspects in the scientific process to be considered. Can we be open in project planning, letting people team up whenever they want in the process? Can we be open in planning our big investments, building our telescopes, our accelerators, and our big projects? Can we be open in data analysis even before we have published the data, and open in data use too? I think so. Our science will benefit from openness in all these respects.

We can learn from history that much of science is primarily not driven by scientists, but by society. However, almost all science is *also* science driven. There are good reasons, both from a scientific point of view and from a societal one, for promoting better science by being open. We also can learn from examples of several of the presentations during this symposium that individuals matter.

By opening up internationally together, we actually can provide something even more important than pure science to the world—namely, demonstrate that together we can do very difficult things. Previously, we have mostly demonstrated this by way of war operations in big international collaborations, but here we can do it with more lasting value. Can we afford to continue losing more than half of all human capacity that would wish and be able to contribute important scientific achievements? Of course not. That is the basic motivation for all these endeavors.

Let me end with a quote that very appropriately is presented as an inscription on the Keck Building, 500 Fifth Street, N.W., where this meeting takes place:

> *The right to search for truth implies also a duty; one must not conceal what one recognized to be true.*
>
> A. Einstein

27. Improving Data Access and Use for Sustainable Development in the South

Daniel Schaffer
Academy of Sciences for the Developing World, Italy

I am speaking on behalf of TWAS, the Academy of Sciences for the Developing World. I will explain what TWAS is, in brief, later in the presentation. At the outset, my talk will focus on both scientific information and scientific data. I will speak about key issues raised during this 2-day symposium, but with a TWAS twist and a particular focus on broad-based problems faced by scientists in the developing world who are seeking to access scientific information and scientific data.

A world-class technical library can be found on the campus of the Abdus Salam International Center for Theoretical Physics (ICTP) in Trieste, where TWAS is also located. When ICTP was created in the 1960s, the library was the centerpiece of the enterprise. Scientists from across the developing world would come to ICTP for the library, since it was unlikely they could get the information at institutions in their home countries. Of course, they would also come to participate in ICTP's research and training activities.

The library is still there, but scientists can now often access much of the library's material from anywhere. The shift that has taken place at ICTP, I believe, is symbolic of the broad changes that have occurred in scientific information and data access across the world.

My presentation will be framed by the United Nations' Universal Declaration of Human Rights, which states that everyone has the right to share in scientific advancement and its benefits. The members of the United Nations signed this declaration in 1947. To that end, there is a long-standing principled foundation to the quest for free and open access to information and data.

As we all know, the number of scholarly and scientific publications are increasing at a rapid rate. It is estimated that there are 25,000 to 50,000 scholarly publications worldwide. Some 2.5 million scholarly articles are published each year. It is estimated that the output is doubling every 15 years. Scientific information is experiencing its greatest transformation since the advent of the printing press more than five centuries ago, and what is true of scientific information is equally true of scientific data. In 2010 it is estimated that the world generated 1,250 exabytes of data. If you placed that data inside a conventional compact disk (CD) and you stacked those CDs one atop another, the stack would rise 3.75 million kilometers, a distance equal to five times to the moon and back. We are generating additional information at such a rapid pace, this year alone we will be producing 1,800 exabytes of data. That means there would be several more stacks of CDs rising to incredible heights.

Some of the problems in communications exist despite the enormous flow of information and data, and some exist because of the endless flow of information and data. For example, there are still some fundamental obstacles that stand in the way of the use, interpretation, and exchange of information and data. Many of these challenges have been mentioned over the course of the past day and a half. Some are universal. The data deluge itself presents an enormous data management problem. Security issues exist at both national and international levels. There are also privacy issues, particularly related to data on public health and medical research. What has not been mentioned extensively here is the reluctance to share data. As we all know, the international scientific community is based on competition and individual accomplishment. That often leads to reluctance on the part of scientists to share data.

Some of the challenges are particular to the developing world. As we heard in the first session today, in poor countries, there is often poor Internet access, limited access to computers themselves, and low bandwidth. It is much less of a problem than it was 10 or even 5 years ago, but it still persists in parts of the developing world. There is also limited access to scientific literature. Historically this has been due to the cost of subscriptions, but we have shifted the burden from institutions to individuals through open access, which has admittedly been an important engine for the spread of scientific information and advances in scientific capacity in the developing world. Nevertheless, open access presents some problems for individual scientists, particularly young scientists.

Last year, TWAS held a conference in Egypt in partnership with the New Alexandria Library on scientific publications. Many of the young scientists who participated in the conference complained about open access because they simply did not have the resources to participate. Charges of $1,000–$1,500 U.S. dollars in author fees to publish in an open-access journal are beyond their means. Indeed it exceeds their yearly salaries. As a result, they do have problems with contributing to and accessing open-access materials, despite all of the benefits that open access provides.

Additionally, there is inadequate training for gathering and interpreting data, poorly equipped laboratories, excessive teaching responsibilities that distract researchers from their research, limited career opportunities that dampen enthusiasm for research, and a general lack of funding. TWAS did a survey about 2 years ago asking young scientists in the developing world, particularly poorer countries in the South, what their research environment was like and what kinds of issues they confronted. We received an e-mail from one scientist in Nigeria, where she had a list of problems with which we are all familiar. I would like to point to the opening and closing paragraphs of her e-mail to highlight the realities those scientists in her circumstances face. She wrote: "For the past 3 hours, I have been trying to reply to your email, but the power has been going off and on every minute." And the last sentence reads: "I cut my discussion short because I need to send this message now before the power goes off." That is the reality that many young scientists face in poor developing countries.

Despite these challenges, there are some—in fact, many—encouraging developments. Two reports published recently indicated that the trends are positive for capacity building and access to scientific publications and scientific data collection in the developing world. These reports are the UN Educational, Scientific and Cultural Organization's (UNESCO) *World Science Report*[5], published in 2010, and the Royal Society's *Knowledge, Networks and Nations*[6], published in March 2011.

The UNESCO report indicated that over the past decade, there has been a substantial increase (from 30 percent in 2002 to 38 percent today) in the number of publications by scientists from the South published in peer-reviewed scientific journals. Yet much of this growth has been due to a very small number of countries. According to surveys done by TWAS, six countries (China, India, South Korea, Brazil, Taiwan, and Turkey) in the global South are responsible for 75 percent of the publications that are being produced in the developing world.

We should note that some of these countries are no longer considered developing (e.g., China is both a developing and a developed country, and South Korea is defined as a high-income developed country in most economic surveys and reports). China, in fact, is now playing a role in the South similar to that played by the United States in the world, in the sense that it is producing 25 percent to 30 percent of all the journal publications in the developing world.

According to the Royal Society's *Knowledge, Networks and Nations*[7], China is on course to overtake the United States in scientific output, possibly as soon as 2013, which is far earlier than expected. Leaving aside the question of impact and overall quality, for sheer quantity, within the next 2 or 3 years, Chinese

---

[5] Available at http://www.unesco.org/new/en/natural-sciences/science-technology/prospective-studies/unesco-science-report/
[6] Available at http://royalsociety.org/policy/projects/knowledge-networks-nations/report/
[7] Available at http://royalsociety.org/policy/projects/knowledge-networks-nations/report/

scientists will likely outpace scientists in the United States in the number of scientific publications they publish.

So, on one side of the spectrum, we have six "developing" countries that are responsible for three-quarters of scientific publications in the developing world. On the other side of the spectrum, we have a group of 80 developing countries that produce very small quantities of scientific information. These countries are home to 1.6 billion people, 25 percent of the world's population. They are responsible for less than 1 percent of the world's scientific publications. Many of these countries are in sub-Saharan Africa.

We have heard examples of superior science being done in the South, and this is undoubtedly true, but the aggregate figures indicate that there is also a growing gap in scientific publications between countries such as China and India, which are progressing rapidly, and others that are lagging farther and farther behind. From TWAS's perspective, one of the key questions about scientific information and scientific data is, How do you deal with these two divergent trends—a narrowing North-South divide that is being matched by a widening South-South divide?

Let me spend a few minutes now talking about TWAS, the Academy of Sciences for the Developing World. Abdus Salam, the Nobel laureate from Pakistan, founded TWAS in 1983 in Trieste, Italy. The secretary general of the United Nations inaugurated the Academy in 1985. It operates under the administrative umbrella of UNESCO. It began with 40 members. It now has 995 members from nearly 100 developing countries: 853 fellows in 74 countries in the South, 142 associate fellows in 17 countries in the North. Fifteen Nobel laureates are TWAS members.

Some of the Academy members may be familiar to you. There is Atta-ur-Rahman, who spoke to the participants at this meeting yesterday from Pakistan via cyberspace. There is Mohamed Hassan, who just stepped down as the TWAS executive director, and the new executive director who was previously on the council, Romain Murenzi. He more recently worked with the American Association for the Advancement of Science in Washington, D.C.

The objectives of TWAS are to:
- Promote excellence in scientific research in developing countries;
- Strengthen South-South collaboration;
- Encourage South-North cooperation between individuals and centers of excellence;
- Respond to needs of young scientists working under unfavorable conditions; and
- Engage in the dissemination of scientific information and sharing of innovative experiences.

Our activities include sponsoring a South-South postgraduate and postdoctoral training fellowship program that we conduct in partnership with such large and increasingly successful developing countries as Brazil, China, India, Kenya, Malaysia, Mexico, and Thailand. We award 175 fellowships per year. Developing countries provide the funding for local expenses and tuition. TWAS provides the funding for transportation to enable these young scientists from the poor developing countries to go to centers of excellence and universities in the host countries. We also have a research grants program for individual scientists, comprising relatively small grants of $15,000, largely used for purchasing equipment and supplies. Despite their modest size, these grants have a great deal of credibility, and they are well known among scientists in the developing world.

We also support institutions. We have a grants program for research groups in poor developing countries. We provide the groups with $30,000 a year over a 3-year period (subject to an annual performance review). These groups have shown much fortitude, ingenuity, and progress in doing research under very difficult conditions, and this money can make a big difference in the quality of their research going forward.

Furthermore, in recognition of the growing capacity of scientific expertise and excellence in the South, we have established, over the course of the past decades, five regional offices. The goal is to develop these regional offices into mini TWAS's that can address within their own regions many of the same issues that TWAS does across the South.

Given what I have said about TWAS, and given the trend toward a widening South-South gap in scientific capacity, how can we manage the growth of data and information for the benefit of all developing countries? It is a key question for TWAS and a key question for most of the members of this audience.

I am going to make a number of recommendations. The good news is these are not radically new ideas. Many of the recommendations have been discussed here. In fact, the recommendations are largely in line with many of the activities, programs, and initiatives that you are involved in. The recommendations also represent a strategy, in the TWAS context, to make them more encompassing so that they do not focus solely on successful developing countries at the expense of developing countries that are not fully participating in international science. I therefore support the following recommendations:

- Access and strategies that provide reduced rates for journals for scientists from poor developing countries. There are a number of initiatives already in play, and they should be supported by institutions like TWAS and institutions that you belong to in order to expand their impact to include the 80 science-poor countries that TWAS has identified.
- Efforts to expand bandwidth and information and communication technology infrastructure.
- Greater participation of scientists in developing countries in international projects. This is happening, but it needs to happen on a broader and more extensive scale across the developing world.
- Strategies for improving the management of indigenous databases.
- The quality and availability of journals and data information produced in the South. The International Network for the Availability of Scientific Publications (INASP) was mentioned in several of the talks. There is also SciDev.Net, which provides extensive news coverage about science and development in the developing world.
- South-South data collection and exchange. Again, there are a growing number of examples, such as the Chinese-Brazil Earth Resources Satellite Program.
- Regional repositories and mirrors in the South. Examples include the National Science Information Center in Karachi and the New Alexandria Library in Egypt.
- More expansive global discussions on data management and use, incorporating the South's viewpoint not just on issues related to training for data acquisition and interpretation, but also on issues related to broader policy and ethical concerns. In fact, in doing reading for this presentation, I noticed that the larger policy and ethical issues are dominated by Northern voices. We need Southern voices to be part of these discussions.
- Best practices in data management in the South. There is India's open-source drug-discovery program among others, and at the ministerial level there is the India-Brazil-South Africa forum.

All of these modest recommendations, which are largely based on existing activities, experiences, and initiatives that have been mentioned at this meeting, can play a critical role in fulfilling the principles and goals articulated in the Universal Declaration of Human Rights with its lofty assertion that everyone has the right to share in scientific advancement and its benefits.

They could also play a critical role in fulfilling the grand vision of Abdus Salam, creator of ICTP and TWAS, who often said in his writings and his speeches, "Scientific thought is the common heritage of mankind."

28. How to Improve Data Access and Use: An Industry Perspective

John Rumble
Information International Associates, United States

The focus of my talk is on how to improve data access and use from an industrial perspective. I work for a private company now, but I worked for the National Institute of Science and Technology (NIST) for many years. During that time, one of my main responsibilities was to run the Standard Reference Data Program at NIST's scientific and technical laboratories. Many of those data programs are run in cooperation and partnership with, as well as with direct funding from, industry. A lot of my perspective arises from my experience in working with various types of industries—from the biotechnology industry hoping to capitalize on genomics, to the concrete industry hoping to be able to build and pave roads in subfreezing weather, to the aircraft industry hoping to take advantage of new composite materials instead of good old-fashioned aluminum.

There are a lot of common features in the way that industry approaches, accesses, uses, and supports data that are useful to it. It is important then to understand how industry looks upon publicly funded data. Just to give you the point of this in advance, industry strongly applauds the open access to publicly funded data because industry is in the best position to take advantage of data from an exploitation point of view, not for scientific credit, but for revenue credit. The more data they have access to, the more comfortable a company is that it is doing things in the best possible way.

It is always useful to look at the life cycle of data. When we talk about access to data, we are really talking about one part of a multistep, continuous process. It starts with measurements and works its way around to having data resources being available and then having people use them. Then, from the use of these data resources, new needs are developed that in turn generate new measurements. This is how the data life cycle process continues.

I want us to think about this particular life cycle from the industrial perspective to see where the interactions of industry take place within this life cycle and to understand some of the places where industry can both help and take advantage of the accessibility of data. Many people do not really realize that industry plays a major role today in a lot of large scale data collection efforts, whether it be the large-scale manufacturer of small instruments that are used over and over again, such as mass spectrometers, genome sequencers, and crystallographic structure machines or all the way up to the big new colliders that have been built with a lot of cooperation from industry. Industry does play a large role in generating data, even publicly funded data, through this support of the scientific effort and through instrumentation. They are looking to make money, but I think it is important to realize that a lot of the sources of publicly funded data resources that have been built in recent years have come as a direct result of industry involvement in this part of the data cycle.

There are situations where industry gets very interested in the development of and access to data resources. Sometimes it comes in terms of direct financial support. Good examples are some of the genomic and proteomic databases that are being built where industrial firms are directly doing measurements and contributing them to larger scale data resources. Strategic moral support in the protein data bank is a good example of where industry does not necessarily provide direct support, but they have strongly encouraged the U.S. government to continue long-term support of that resource.

The knowledge support services that some of the best data scientists have developed in terms of designing data architecture, data resources, and web services came from IBM, Microsoft, and now Google. In other situations, where there are not particularly large sets of tools or methodologies to exploit data, industry does a lot of development in terms of data mining, visualization, and what I call "value added". This is where industry will take some publicly available data, add value to it, and then sell it. An example is the NIST mass spectral database, which contains several hundred thousand mass spectra. NIST distributes this database with a nice interface, but some of the machine manufacturers have taken that product and added an analytical package on top of it, thus enhancing the value of that publicly funded data resource by making it a more important scientific and technological tool.

Finally, one other way that industry very importantly involves itself in the data life cycle is by articulating new needs. Those of you from university research programs are well aware that industry has started many academic-industry joint programs. These programs have been started because industry realizes it has need for new scientific knowledge and measurements that it cannot satisfy itself.

Another important main point has to do with how industry accesses data and when it accesses them, why it accesses them, and the economic implications of its access. Almost any product and service that is available commercially starts as a concept. It might simply be one entrepreneur sitting at a restaurant sketching something, or the result of a long process by a complex design team. What is important to understand is when industry makes money. Industry does not make money from concepts or rough designs; it makes money from selling things.

The main point I want to make here is that the industry's willingness to pay for data is directly related to how close in time the use of data are to the sale of a product or service. Data used years before a product is released and sold is not "valued" as highly as are data whose use has immediate impact. An example of the latter is analytical chemistry data that are used to identify an unknown substance that has affected a manufacturing process. Because the impact is immediate, companies are willing to pay many thousands of dollars for data. In contrast, companies are less willing to pay for material property data used early in a multi-year design database.

The picture I am trying to paint here is that industry needs to have access to all kinds of different data to support product development and services so they can make money. Data comes in lots of different flavors and has lots of different uses. Data also has differing impacts on the ability of industry to make money from data use depending on when that use occurs. Industry is willing to pay for data when it perceives the value of those data in helping them make more money.

What really incentivizes industry to participate more willingly in scientific data activities? How do we improve access to data? There are situations where industry's participation in scientific data activities is extremely important to the progress of those scientific data activities. Obviously the first example that comes to mind is genomics, but genetically modified foods are another case. Other examples are pharmaceutical development, and aircraft manufacturing.

The primary motivation is the potential for increased revenue. Rarely do companies act out of goodwill. If we want industry participation in more public scientific data activities, it is perfectly acceptable to allow them to demand a business case and for you to provide that business case to them, because that is how industry makes decisions. There are subsidiary reasons, such as intellectual property rights, which eventually translate into increased revenue or increased market share. A company will support a data project if it leads to development of a new product, for example, a specific new pharmaceutical, such as a new drug for diabetes, which in turn will lead to increased revenue.

There are other less tangible ways that are also useful to think about. One is that there are many smart people in industry, and they know that new science is going to create new industry and new products. They are not sure how, and if they participate in the fundamental research that develops these scientific disciplines, they will have an opportunity to perhaps get the insights that will lead to revenue. The professional societies in the United States have incentivized industry to cooperate together. A lot of the large-scale data programs are funded by industry through professional society programs.

I would like to end by saying that contrary to what most people think, industry really supports open access to publicly funded data. From their perspective, they are often in the best position to exploit them for whatever reasons they want to. If industry have data that it generated, there are many mechanisms to keep that proprietary data to themselves. As I already indicated, however, there are instances when they do generate data or do help the generation of data, which contributes considerably to publicly funded data resources.

29. Production and Access to Scientific Data in Africa: A Framework for Improving the Contribution of Research Institutions

Hilary I. Inyang
African Continental University System Initiative[8]
University of North Carolina, Charlotte, United States

INTRODUCTION

To begin, I would like to describe the role of universities during three eras in Africa. The first was the pre-independence era in Africa when universities were engaged very deeply in independence movements, protests, and so forth. In Mozambique, for example, university-based intellectuals were at the forefront of protests to gain independence from the Portuguese government. At that time, African universities were at the vanguard of diverse indigenous groups that rationalized the need for the independence of African countries on the basis of the human rights to freedom and self-governance. They were not really deep contributors of data or other forms of information to economic development initiatives and governance of their countries. That role was played by colonial governments directed from Europe. This circumstance was prevalent in Africa until the late 1950s.

The 1960s, when the wind of change blew across Africa bringing with it independence, was another significant period with respect to the availability of data in Africa for national economic development programs of newly independent nations. The first set of post-independence leaders in Africa were statesmen, exemplified by Dr. Kwame Nkrumah of Ghana, Dr. Kenneth Kaunda of Zambia, Dr. Siaka Stevens of Sierra Leone, Dr. Leopold Sedhar Senghor of Senegal, Dr. Julius Nyerere of Tanzania, and Dr. Nnamdi Azikiwe of Nigeria. Most of them were western-educated and valued good education, as well as the need to develop and use data for their governments' economic planning and governance. The high educational standards of the era are highlighted by the fact that despite the existence of very few research institutes within the continent at that time, colleges were able to engage in intellectual enquiry to generate data and build human capacity on analytical aspects of economic program planning and implementation. However, during that era, universities were strong in the liberal arts, but not so much in the sciences, because they had not yet developed the infrastructure that would have made them competitive with those in global science.

Unfortunately, following their initial interest in democracy and the utility of knowledge systems, including data in the sustainable development of their countries, most of the early leaders overstayed their welcome in leadership positions and sometimes, turned to autocracy. The ultimate result was a wave of military *coup d'états* across Africa in the 1970s and 1980s, that derailed most of the long-term educational and research initiatives that would have institutionalized the generation of data for economic planning and project implementation in Africa. Most of the new military leaders installed autocracies that devalued knowledge systems and operated through edicts without regard to scientific facts and data that were not in support of their decisions. Intellectuals were often prosecuted and many were driven into exile.

Since the 1990s, there has been a continuing diminution of autocracy in Africa. The need for national planning is recognized. Most African countries have 5-year national development programs, and they are beginning to realize that this is the time to extract the intellect of their people and invest such intellect in development efforts. This era has also witnessed the emergence of continental knowledge systems consortia, professional organizations and academic institutions that target the generation of knowledge and information. Examples are: the four-campus African University of Science and Technology, the Pan African University System, and continental professional societies in virtually every major scientific field.

---

[8] Former president of the African Continental University System Initiative.

Organizations such as International Council for Science – Regional Office for Africa (ICSU-ROA) and UNESCO have embarked on science support activities that will also generate data. An example is the ICSU-ROA Science Plans in many thematic areas that are critical to Africa's sustainable development. Requirements for improvements of data generation, management, and utilization systems in Africa should be viewed within the context described above.

THE ROLE OF KNOWLEDGE GENERATION INSTITUTIONS IN DATA PRODUCTION AND MANAGEMENT

What are the typical roles of universities and other institutions that are engaged in research? I see such roles to be the following:

- Supplier of options for sustainable development;
- Producer of data for decision support systems;
- Developer of human resources and capacity;
- Creator of innovative ideas and products; and
- Guardian of rationality and human rights.

The last one often puts universities in conflict with political authorities. Every dictator that shows up wants to imprison journalists and professors. Let me start by addressing some problems. There are some very large projects in Africa. Development banks such as the World Bank and the African Development Bank sponsor most of these projects. Very large companies, more recently large Chinese companies, also sponsor such big projects. A main problem with these initiatives (e.g., building a dam or developing a big mining facility) is that there are no clear requirements regarding post-project use of the data and information that they generate for other sustainable development programs of the host countries. Then, of course, there is the issue of sensitive information and how to deal with it.

Also, there is the need for coordination and collaboration. For example, most of the African countries have declared a set of activities to achieve the Millennium Development Goals, which are very specific. All of these activities will require data and other types of information for planning and implementation of projects. This is why I see that there is a need for a systematic relationship between data access programs and these efforts.

CAPACITY LIMITATIONS OF AFRICAN COUNTRIES

Furthermore, there is the issue of the resource and human capacity gaps among African countries. It is not that information is sparse in all parts of Africa on every issue. In some countries like South Africa, Tunisia and Algeria, there is a lot of information, but some countries in West Africa and Central Africa do not have adequate facilities and capabilities for research information generation on critical issues in many economic sectors. About 30 percent of the annual budget of some of these poor countries comes as foreign aid. They have other things that they consider more immediate and more expedient than developing research facilities and data management systems.

Additionally, we should always remember that Africa has a number of official languages. So, it is very difficult to do things on a region-wide basis. Even in West Africa, there are 5 Anglophone and 9 Francophone countries. In Central Africa, French and Portuguese are used, and in Equatorial Guinea, Spanish is used. Translation of documents and real-time oral speeches can be very expensive.

The level of investment in data generation activities is very limited in African countries. In fact, the entire African investment in this area is less than that of Israel. Currently, there are about 500 science parks worldwide, but less than 2 percent of those parks are in Africa. If those parks are absent, how will data be transferred or managed as we do here in places like the Research Triangle Park in North Carolina or Silicon Valley in California? There should be a clear model for how universities can interrelate with other organizations to promote economic activities and progress. This has to be promoted through, for example, the activities of organizations like the World Bank, the African Development Bank, UNESCO, ICSU, the United Nations Economic Commission for Africa, the African Union, the Regional Economic Blocs, and international aid organizations, as well as the African governments themselves.

PROPOSED SOLUTIONS
What are the solutions? We need to establish an African Research Foundation. Most of the efforts and money that have been spent would be better spent on the development of this critical entity. Even in Europe, where there are very strong national science or research foundations, there is still the superposition of the European Science Foundation on a continental basis, on those national capabilities. I think it is very critical to do this in Africa as well, more so when the lack of effective national research agencies is pervasive in the continent, except of course, in South Africa. An endowment fund should be attached to such an endeavor, because if there is a research foundation, it would be ineffective, if it does not have the funds to deal with issues.

As in most countries, universities in Africa emphasize three broad research disciplines: the social sciences and humanities; engineering, technology and computing sciences; and physical, biological, ecological, and other natural sciences. In parallel, governments employ a variety of policy tools: market incentives, risk communication, technology deployment, public education, and so forth. Unless there is an interplay between these academic approaches and government policies and initiatives, it is very difficult to maximize the value of data, information, science, and public policy.

I would like to address some important developments in Africa in the information technology area. There are many optical cable systems that are being implemented now in Africa. These cable systems are improving internet access in Africa. For example, the West African Cable System is reducing the cost of Internet connectivity, including access and data transfer, and mobile systems. As a complement to this development, there are about 10 satellite systems now covering various regions, including sub-Saharan Africa and the Arab countries in the north of Africa.

CONCLUSIONS
Let me conclude by stating that the formation and operation of an African Research Foundation would greatly enhance the generation of data and other types of information for regional sustainable development. Such a Foundation should have the following mandate:
- To support research for production of information, including data, for use in African sustainable development programs;

- To provide an opportunity for engagement of African and other experts and the development of African talent (in Africa and the diaspora) on science and technology research on critical issues that affect Africa; and

- To catalyze science and technology-based African entrepreneurship and improve the infrastructure for access to data in locations within and outside Africa.

If these recommendations are implemented, Africa will increase its contribution of knowledge, not only to the improvement of the quality of life in the continent, but to global sustainable development. It is a fact of history that the continent has a heritage of contributions to human development and on intellectual matters throughout the centuries.

30. The ICSU World Data System

Yasuhiro Murayama
National Institute of Information and Communication Technology, Japan

My talk will include both international activities related to ICSU's World Data System, as well as some examples of scientific database and data sharing activities related to developing countries, especially in Asian countries. I hope that the specific examples I will share with you can inspire you to promote more open access data for social benefits.

About 50 years ago, when the former "World Data Center" (WDC) system was established, the main objective was data preservation primarily for printed data sheets and films, before digital data become predominant with state-of-art technology of high-speed Internet, and huge data storage in computer servers. Those were valuable efforts to keep scientific data indispensible. Today, we get more and more data in digital form, and the size of the holdings is increasing rapidly. In Japan, other Asian countries, the United States, and Europe, we are expanding our data strategies, and scientists are concerned with how to handle the data influx. Also with digital data, the future interoperability for world-wide data centers connecting to each other can be envisioned, which was not available for data stored on paper and film.

In this context, the International Council for Science (ICSU) established the World Data System (WDS) in 2008. ICSU envisioned a global data system that would:
- Emphasize the critical importance of data in global science activities;
- Further ICSU strategic scientific outcomes by addressing pressing societal needs (e.g., sustainable development and the digital divide);
- Highlight the positive impact of universal and equitable access to data and information;
- Support services for long-term stewardship of data and information; and
- Promote and support data publication and citation.

I should mention that CODATA has also participated in such international data sharing efforts, and we are hoping for more future collaboration with CODATA. Other collaborators come from the disaster research area. We also have collaborations with international scientific unions, a number of United Nations agencies, and additional Asian and Japanese unions.

To fulfill this vision and to facilitate collaboration, I am now in process of establishing a new International Programme Office (IPO) for ICSU-WDS this year, as its acting director. The IPO is to be hosted by the National Institute of Information and Communication Technology (NICT) in Japan (NB: the internationally selected Executive Director was appointed in March 2012, and the official opening ceremony was held in Tokyo, in May 2012).

Next, I would like to address several scientific data activities carried out by myself and my institute. We have a monitoring network for solar and space science, and space weather observational data. These kind of data are important in social activities too; for example, the ionosphere is the biggest delay factor in radio navigation signals for the Global Positioning System (GPS). This is very crucial in positioning aircraft and in future precise applications.

If we want to introduce GPS-guided aircraft navigation, the networks should extend into the south Asian countries and be based upon sound science where data are shared together with those countries. Many cities also provide this kind of distributed system and use high-speed internet to link with several Japanese universities.

Future directions include scientific data cloud services, or creating a digital Earth capability that combines observational data and simulation data with informatics technologies. Better management of the data can be enabled through high-speed research network experiments, such as the Asian-Pacific Advanced Network (APAN). This kind of infrastructure enables Japanese researchers to access data from various countries in eastern and southern Asia and vice versa.

A final example focuses on university groups in Japan. They are now designing a metadata system for space and atmospheric, and interdisciplinary sciences. An improved metadata system enables more information exchanges to promote data usage and more scientific research. Universities also have various data observing networks. Some are radars in Arctic regions, while others are radars or magnetometers in Asian countries or in Antarctic stations. The starting point is a good metadata system for data in the archives.

I would like to conclude by noting that in September 2011 we will have the first World Data System conference in Kyoto, together with the WDS scientific committee meeting.

31. Libraries and Improving Data Access and Use in Developing Regions

Stephen Griffin
National Science Foundation, United States

I am going to explore some of the potential roles for libraries in improving access, use and sharing of large stores of data in support of e-Science and other data-intensive scholarly inquiry. The focus will be on their application in developing countries.

Contemporary research and scholarship is increasingly characterized by the use of large-scale datasets and computationally intensive tasks. Vast amounts of data are used by scientists to better map the cosmos, build more accurate earth system models, examine in finer details the structures of living organisms, and gain insights into the behaviors of societies and individuals in a complex world increasingly dependent upon information technology.

Significantly, more humanists are rapidly integrating newly digitized corpora, digital surrogates of cultural heritage artifacts, and historical, spatial, and temporal indexed data into their scholarly endeavors. The datasets are huge by any measure. Petabyte-scale datasets are not uncommon. More raw data are being produced today than can be physically stored, and this disparity will almost certainly increase very rapidly.

Much has been accomplished by the libraries and information sciences communities over the past decade to establish basic principles, standards, object representations, descriptive metadata, and reference models needed to achieve interoperability, scalability, access, and long-term preservation and archiving of digital content.

Already, libraries have undergone significant transformation by converting collections and holdings to digital forms in prescribed formats with appropriate descriptive metadata. In addition, they have dealt with a deluge of new "born digital" data from a variety of sources. Libraries now play a central role in providing enhanced access to very large digital collections across many topic domains for a wide variety of users, but the prospect of libraries taking responsibility for large-scale raw datasets and a multitude of derivative forms of data associated with e-Science and data-intensive scholarship was not seriously considered until more recently. As a result, there is lively debate and controversy over how libraries, and particularly research libraries, should participate in providing tools and infrastructure in support of data-intensive research.

The challenges are daunting. In addition to tasks inherent in the life cycle of data and information, there are other tasks requiring new technologies and new expertise and a broader spectrum of user services. This points to the question of how graduate schools of library and information science should prepare students for the realities that await them in a digital world of knowledge resources.

For developing countries, all of that applies, but there are additional barriers that are not present in more-developed regions of the world. At the same time, perhaps there are unique opportunities that we can discover, as in some respect they are beginning at a different starting point in a long evolutionary process and may not be burdened by entrenched practices.

There are several points in Figure 31-1 that depicts the rapid transition from storing information on analog media to digital media. It is impressive how quickly it happened, and perhaps a bit disconcerting, too. Also, it should be noted that scientific data only represents a very small proportion of the data being produced, and all data of value require careful management to ensure reusability. Of equal importance is that the various media on which the data are stored deteriorates over time and will likely become unreadable within a few decades. This implies that policies for physical migration be established and followed.

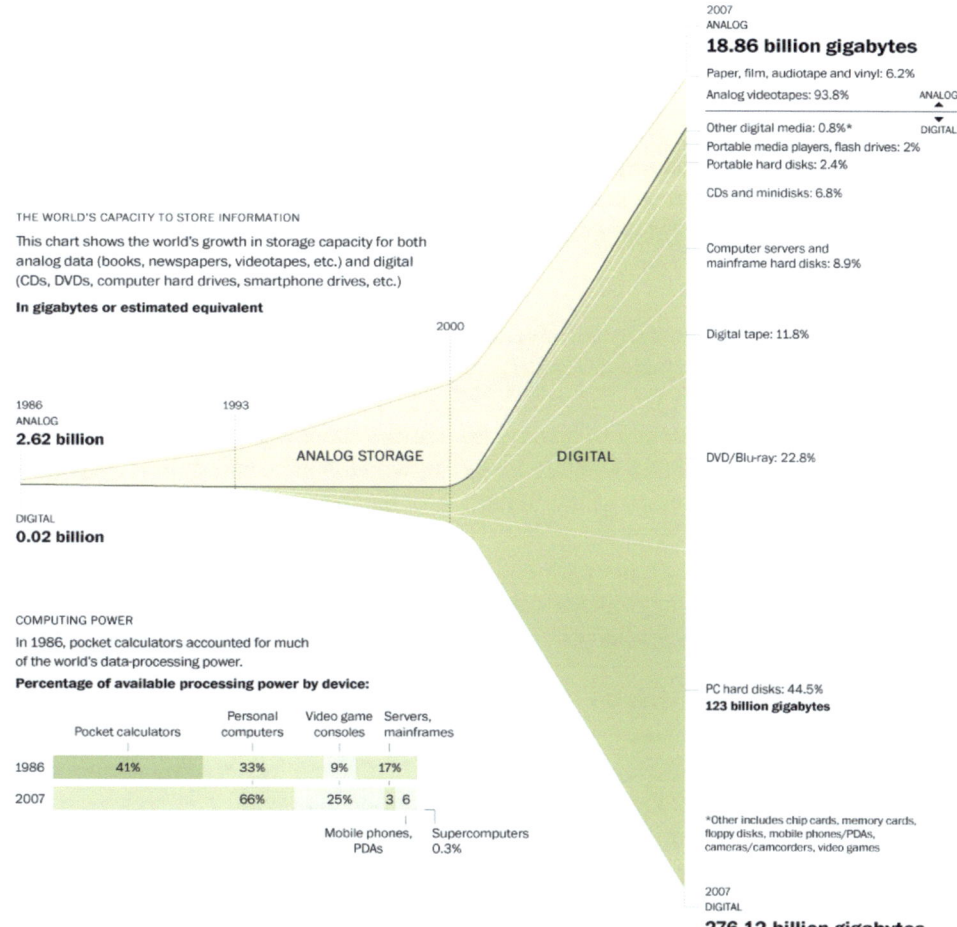

FIGURE 31-1 *The World's Capacity to Store Information*
Credit: Hilbert, M. & López, P. (February 11, 2011). The World's Technological Capacity to Store, Communicate, and Compute Information. *Science Magazine, 332*(6025), 60-65.

Contemporary data-centered research and scholarship can be categorized based on the types of data associated with the research activity:

e-Science: A natural evolution of computational science that involves massive simulations of phenomena of scientific interest too large or small, too fast or slow, or too complex to explore in a research laboratory. *e*-Science is computationally-intensive and frequently involves the use of distributed network environments and grid computing.

Data-Driven Science: A rapidly growing set of applications in which analysis of large amounts of experimental data drives the overall research. The sources of data are often high-throughput digital instruments and recording devices; for example, sophisticated astronomical instruments, particle accelerators, environmental sensors, medical diagnostic equipment, and many others. The petabytes of data that may result frequently require significant computation to yield the basic data for analysis. Data-driven science is a more recent paradigm primarily dependent upon new forms of data analytics.

Data-Intensive Research and Scholarship: This includes efforts based in part on the exploration, manipulation, and analysis of diverse datasets including born-digital data, data resulting from the continuing digitization of large analog collections, digital representations of physical artifacts and complex higher-order derived data constructs. Collaboration within and across disciplines may take a key role. Interoperable and highly contextualized datasets are often essential to success.

Data-intensive research is being broadly pursued in many disciplinary domains. Scholars in the humanities and social sciences are finding new problems of compelling interest that both have a strong technology component and are based on analysis of large and heterogeneous datasets. Collaborative efforts that involve technologists and domain researchers are leading to advances in knowledge and new understanding in many areas.

In all of the categories of research described above, data visualization of research findings is common and frequently the most effective way to present results.

Early programs at NSF were instrumental in leading the way to some of the data-intensive research that we are seeing today. The Supercomputer Centers Program, which was begun in the mid-1980s, and the NSFNet program that was originally part of the Supercomputer Centers Program in the 1980s, were very important. The NSFNet was originally envisioned to be a service-type of facility for the supercomputer centers. The idea was that if one built a network connecting large computing facilities around the United States, one could share datasets and aggregate computational cycles. The notion of a network as a means for federated search of data repositories had not yet been seriously discussed.

As the network evolved, it began to connect data stores. New access methods were developed and the idea of digital libraries emerged. I was given the responsibility to be the Program Director for the multi-agency and international Digital Libraries Initiatives (DLI) that began in 1994 and continued as a programmatic entity well into the 2000s. This program is well documented. The broader impact of the program included not only noteworthy achievements in advancing academic research and scholarship, but also created profitable commercial entities as well; Google, Inc. being a prime example. The legacy of the DLI Program continues as new interdisciplinary communities develop and use digital content and advanced information infrastructure and tools to probe new problem domains. The culmination of support for large-scale computing and networking infrastructure was the NSF Cyberinfrastructure program, based on the findings outlined in a report written by a group led by Dan Atkins of the University of Michigan.

I want to focus on some of the disadvantages of researchers in developing countries. While some of the maps we have seen show increased connectivity in developing countries, bandwidth is not sufficient or uniform enough to support the data-intensive scholarship that I have been describing. For a researcher in a less developed or developing country, one of the primary considerations is a sense of isolation. Not only a sense of not being able to gain access to critical resources, but also of being left out of the mainstream of the scientific discourse that characterizes their particular disciplines and their particular research areas.

What is likely to happen? I think that the roles will be enhanced for certain libraries. Certainly the national libraries and the university and research libraries will be expected to carry the load. This is because they are likely to be the best equipped to handle data. They have been doing some of it for many years and know what makes it work. It is also because the scientists in general have very little interest in handling their own data. They want to use them, get their experiment done, and publish their results, but that is it, in most cases. We now are becoming aware that the reuse of data is of critical importance in terms of saving resources.

Finally, there are special topic libraries. These may be libraries of a single individual, but they have an extraordinary wealth of information. Often the special libraries focus on a particular media like photographic libraries, image libraries, film libraries, or audio libraries, and these will become increasingly important as research resources for numerous scholarly domains.

While there is no single designation that is generally accepted to determine whether a country is developing or not, there are some general comments that one can make. Making data useful and facilitating scholarly communication and data curation services are among the most important and the most tractable activities for libraries in developing countries at present. What is more difficult is to assist directly in the establishment of close peer relationships with counterparts in other countries, provide high-bandwidth access to remote and distributed information stores, build advanced content management schemes, create sophisticated applications tools, and provide the level of support and services that libraries in more developing countries are able to provide.

There are some things that these institutions could begin immediately to do that might in fact bring them into the mainstream without the physical infrastructure enjoyed by their counterparts in more economically developed countries:
- Increase staff dialogue in all institutional and technical matters;
- Strive to build a sense of local community and institutional pride (e.g., pursue excellence and recognition in specific areas, create new collections and services, and create and "publish" a new journal highlighting local research);
- Enhance outreach to the larger regional and global communities (e.g., support staff attendance and participation in relevant conferences and professional training events, and establish exchange programs);
- Expand efforts to convert digital holdings to current standards and forms;
- Build new tools and resources; and
- Pursue internal and external support from public and private sources (i.e., UNESCO, USAID, the European Union, and various non-governmental organizations and private foundations).

The emergence of a culture of sharing has accompanied the growth of the Internet and new communities involved in the creation and management of digital content and resources. The digital libraries community has been particularly instrumental. At the same time, there has been a general movement toward openness of internet based content and resources (when appropriate and legal). For example:
- Open data
- Open access journals
- Open repositories and archives
- Open source software
- Open architectures
- Open educational resources
- Open and transparent governance
- Open scholarly and practitioner communities

The Directory of Open Access Journals (DOAJ) is one example. The DOAJ lists more than 8000 journals (as of October 2012), from more than 110 countries in more than 50 languages. Over 4000 journals are providing metadata on the article level, which means that almost 900,000 articles are searchable from DOAJ.[9] There are other global efforts that are the subject of this symposium.

---

[9] See www.doaj.org

Geography is a primary consideration in building high-bandwidth networks. In the case of Africa, many factors influence the cost and effort. The size of Africa alone presents formidable challenges. GÉANT[10], the pan-European data network, has done a great deal of work in connecting various regions of the world. Internet2 has large international partnerships, and beyond their partnerships, they have other networks that are reachable. National Research and Education Networks (NRENs) are an important core element of a country's networking capabilities, as they are most often funded as part of a nation's budget and provide a single, robust connection point to NRENs in the rest of the world. NSF has an international network connections program.

Libraries will continue to play an essential role in the internetwork society. Libraries have managed the collected knowledge of the human record for centuries and will continue to do so. The manner in which this is done for digital content implies substantial and rapid change in established practices. This is being done already and new communities of practice are emerging that will continue to develop the means for confronting the ever-expanding volumes of data being created.

---

[10] See http://www.geant.net/pages/home.aspx

32. Developing a Policy Framework to Open up the Rights to Access and Reuse Research Data for the Next Generation of Researchers

Haswira Nor Mohamad Hashim
Law and Justice Research Centre
Queensland University of Technology, Australia

The Next Generation of Researchers

In order to identify who is the next generation of researchers, we need to distinguish them from the current generation of researchers. For the purpose of this paper, the current generation of researchers is identified not by their age or seniority, but by the research methodology that they adopt in their research projects and the medium of dissemination of their research data and information. Seen from this perspective, the current generation of researchers is a generation that is already taking advantage of information and communication technology by employing e-Science and e-Research methodologies and relying more on research collaborations that require intensive data sharing at both national and international levels. The current generation of researchers disseminates research data through both print and electronic media. Besides print publications and conference proceedings, research data were also published in open-access journals or self-archived in open-access data and repositories.

As for the next generation of researchers, it could be anticipated that they will continue to take advantage of the advancement of information and communication technology and be involved in interdisciplinary and multidisciplinary research collaborations, albeit at greater scale. Like the current generation of researchers, it is predicted that the next generation of researchers will continue to publish and self-archive their research data in open-access journals and open-access repositories as alternatives to conventional methods of dissemination. At first glance, it seems that the next generation of researchers will share many similarities with the current generation of researchers whose roles they will take over in the future.

Despite these similarities, it can be observed that the current generation of researchers has become the victim of a serial crisis arising from a toll-access system widely practiced by journal publishers.[11] Apart from that, the rapid expansion of copyright protection and the introduction of sui generis database rights in the European Union and a few other countries, which took place during the era of the current generation of researchers, has resulted in almost all categories of research data becoming the property of either the researchers or the research institutions that employ the researchers, or the research agencies that fund the research.[12] Research data, which contains a wealth of information is no longer freely accessible.[13]

---

[11] See Open Access to Research Outputs: Final Report to RCUK, 2008, 21, available at <http://www.rcuk.ac.uk/cmsweb/downloads/rcuk/news/oareport.pdf> (accessed March 11, 2010); Science and Technology Committee, 2004, Science and Technology – Tenth Report, U.K. House of Commons, Science and Technology Committee.

[12] The United States has not adopted sui generis protection for databases, in contrast to Europe.

[13] See The Royal Society (U.K.), 2003, Keeping Science Open: The Effects of Intellectual Property Policy on the Conduct of Science, 21, available at <http://royalsociety.org/Report_WF.aspx?pageid=9842&terms=Keeping+science+openpercent3A+the+effects+of+intellectual+property+policy+on+the+conduct+of+science> (accessed February 26, 2010) v.; Paul A. David, 2004, Can "open science" be protected from the evolving regime of IPR protections? 129(March) *Journal of Institutional and Theoretical Economics* 3; Pamela Samuelson, 2003, Mapping the digital public domain: Threats and opportunities, 66(1 & 2) *Law and Contemporary Problems* 170; James Boyle, 2003, Foreword: The opposite of property? 66(1 & 2) *Law and Contemporary Problems* 13; Jerome H. Reichman and Tracy Lewis, 2005, Using liability rules to stimulate local innovation in developing countries: Application to traditional knowledge, in Keith Eugene Maskus and Jerome H. Reichman, eds., *International Public Goods and Transfer of Technology Under a Globalized Intellectual Property Regime* (Cambridge University Press) 340; P. A. Andanda, 2006, Human-tissue-related inventions: Ownership and intellectual property rights in international collaborative research in developing countries, 34 *J Med Ethics* 172; Samuel E. Trosow, 2003, Copyright Protection for Federally Funded Research: Necessary Incentive or Double Subsidy? 18, available at <http://publish.uwo.ca/-strosow/Sabo_Bill_Paper.pdf> (accessed September 12, 2010); Charlotte Hess and Elinor Ostrom, 2003, Ideas, artifacts, and facilities: Information as a common-pool resource, 66 *Law*

The current generation of researchers' access to and reuse of research data are subject to consent or license from the owners or custodians of research data. Further, the emphasis on commercialization of publicly funded research outputs had seen data access and sharing practices replaced by the norms of secrecy, whereby data withholding (including by academic researchers) is widely practiced, resulting in the tragedy of anticommons.[14]

While the current generation of researchers and the next generation of researchers may share many similarities, the current generation of researchers is seen as a generation deprived of their rights to access and reuse research data and scholarly information. It is therefore deemed important to ensure that the next generation of researchers will not suffer the same disadvantages as their predecessors.

The Need to Open up the Rights to Access and Reuse Research Data for the Next Generation of Researchers

It is predicted that the roles that the next generation of researchers are going to play will be far more challenging in understanding global phenomena, solving problems, and improving the human condition, which are the aims of basic and applied research. Global crises such as climate change, rapid population growth, scarcity in natural resources, and natural disasters will be inherited by the next generation of researchers from their predecessors with a greater sense of urgency, as the problems are expected to be more acute. In solving these problems, the next generation of researchers will depend on the latest scientific data in their field, especially in the fields of research where a global picture is necessary for the development of effective international programs to solve transboundary problems such as infectious disease control and monitoring, environmental conservation, and climate change.[15] Without the ability to access and reuse research data, their research projects become meaningless, if not impossible.[16]

It could also be anticipated that the scientific research that will be undertaken by the next generation of researchers is becoming data driven and data intensive at a scale previously unimagined.[17] It is predicted that data-intensive science will emerge as a new paradigm of scientific research. This data-intensive science was described by Jim Gray as the "Fourth Paradigm,"[18] which is a new addition to the previous three scientific paradigms: theory, experimentation, and computational science.[19] The emergence of data-intensive research undertaken by scientists on a global scale requires opening early-stage research data in order to encourage broader participation and accelerate discoveries.[20] In playing their roles as researchers

---

and *Contemporary Problems* 112; Carol M. Rose, 2003, Romans, roads, and romantic creators: Traditions of public property in the information age 66 *Law and Contemporary Problems* 90.

[14] See Bhaven N. Sampat, 2006, Patenting and US academic research in the 20th century: The world before and after Bayh-Dole, 35 *Research Policy* 784; Dirk Czarnitzki, Wolfgang Glanzel, and Katrin Hussinger, 2009, Heterogeneity of patenting activity and its implications for scientific research, 38 *Research Policy* 33; Jerry G. Thursby, Richard Jensen, and Marie G. Thursby, 2001, Objectives, characteristics and outcomes of university licensing: A survey of major U.S. universities, 26 *Journal of Technology Transfer* 59; Peter D. Blumberg, 1996, From "publish or perish" to "profit or perish" : Revenues from university technology transfer and the s. 501(c)(3) tax exemption, 145(1) *The University of Pennsylvania Law Review* 91; Andrew F. Christie et al., 2003, Analysis of the Legal Framework for Patent Ownership in Publicly Funded Research Institutions (Commonwealth of Australia Department of Education, Science and Training) 48.

[15] Barbara Kirsop and Leslie Chan, 2005, Transforming access to research literature for developing countries, 31(4) *Serials Review* 24.

[16] Marjut Salokannel, 2003, Global Public Goods and Private Rights: Scientific Research and Intellectual Property Rights, 11, available at <http://www.iprinfo.com/tiedostot/5icFWowu.pdf> (accessed October 12, 2010).

[17] Michael L. Nelson, 2009, Data-driven science: A new paradigm? 44(4) *Educause* 6.

[18] See Tony Hey, Stewart Tansley, and Kristin Tolle, eds., 2009, *The Fourth Paradigm: Data-Intensive Scientific Discovery* (Microsoft Research, Redmond, Washington).

[19] Michael L. Nelson, Data-driven science, 6; Francine Berman and Henry E. Brady, 2005, Workshop on Cyberinfrastructure for the Social and Behavioral Sciences: Final Report (National Science Foundation) 32.

[20] Anne Fitzgerald and Kylie Pappalardo, 2007, Building the infrastructure for data access and reuse in collaborative research: An analysis of the legal context, in Open Access to Knowledge (OAK) Law Project: Legal Protocols for Copyright Management: Facilitating Open Access to Research at the National and International Levels, 58, available at <http://www.oaklaw.qut.edu.au/files/Data_Report_final_web.pdf> (accessed January 11, 2010).

amidst data-driven and data-intensive science, access to and reuse of research data will be essential for the next generation of researchers.

Besides the emergence of data-intensive science to solve global problems, another reason the next generation of researchers needs to be given the rights to access and reuse research data is to facilitate their participation in international, interdisciplinary, and multidisciplinary research collaborations. As global society turns to science with more and more problems, it will require the collaboration of the next generation of researchers from various disciplines to think about solutions to the same problem.[21] In this regard, science can no longer be managed within the old silo research model, as it requires thousands of collaborators across distinct institutional boundaries.[22] Collaboration among researchers is said to be an interactive process, where two or more researchers, or research organizations, work together toward common objectives by sharing knowledge and research results using collaborative research networking tools.[23] The collaborative research requires the ability to search, access, move, manipulate, and mine research data like never before.[24]

As discussed above, global research collaboration requires the next generation of researchers all around the world to contribute and share their research data and information online. The National Science Foundation report *Revolutionizing Science and Engineering Through Cyberinfrastructure* stated that wider and easier access to reports, raw data, and instruments are needed in order to change science and engineering research.[25] Fry et al. argued that e-Science and e-Research collaborations are not only about access to secondary resources such as scientific articles, or to primary resources such as databanks, but also to openness in tool development and the sharing of software code to extend and modify tools.[26] As most research data are protected under a copyright or database-right legal regime, opening up the rights to access and reuse research data will form a big part of the emerging infrastructure for globally organized collaborative research activities.

The need to open up the rights to access and reuse research data also arises from conventional research libraries and the traditional publishing system not being well suited to the need of the next generation of researchers, who will require not only access to the latest research data but also the ability to communicate their research results as quickly as possible. While access to and reuse of research data are highly in demand, research libraries and journal publishers have not been typically responsible to archive research data, so the role of the library and publishers as data steward is largely absent.[27] In contrast, opening up access to and reuse of research data facilitates the collaboration of various research efforts, which in a closed-access world are circumscribed by conventional definitions of topic, field, or discipline and are isolated from one another in discrete families of journals.[28] With proper infrastructure such as open-access repositories, researchers in different domains can collaborate on the same dataset, and use, reuse, and combine data, which increases productivity and reveals new insights.[29] Based on the above

---

[21] Francine Berman and Henry E. Brady, 2005, Workshop on Cyberinfrastructure for the Social and Behavioral Sciences: Final Report (National Science Foundation) 57; Committee on National Statistics,1985, Issues and recommendations, in Stephen E. Fienberg, Margaret E. Martin, and Miron L. Straf, eds., *Sharing Research Data* (National Academy Press, Washington, DC) 24, 25.
[22] Yochai Benkler, 2008, The university in the networked economy and society (November/December) *Educause* 65.
[23] David Bicknell, 2009, Collaboration drives innovation (Apr 28–May 4) *Computer Weekly* 14.
[24] Tony Hey and Anne E. Trefethen, 2005, Cyberinfrastructure for e-Science, 308 *Science Mag* 818.
[25] Daniel E. Atkins et al., 2003, Revolutionizing Science and Engineering Through Cyberinfrastructure: Report of the National Science Foundation Blue-Ribbon Advisory Panel on Cyberinfrastructure (National Science Foundation) 17, 28.
[26] Jenny Fry, Ralph Schroeder, and Matthijs des Besten, 2009, Open science in e-Science: Contingency or policy? 65(1) *Journal of Documentation* 7.
[27] Neil Rambo, 2009, E-science and biomedical libraries, 97(3) *J Med Libr Assoc* 160.

[28] Alma Swan, 2007, Open access and the progress of science, 95(May-June) *American Scientist* 200.
[29] Kostas Glinos, 2010, Report of the High Level Expert Group on Scientific Data: "Riding the Wave: How Europe Can Gain From the Rising Tide of Scientific Data – A Vision for 2030" (European Commission.

arguments, the need to open up the rights to access and reuse research data for the next generation of researchers is therefore justified.

The Next Generation of Researchers' Responsibilities as Providers and Users of Research Data

As discussed in the previous section, current developments point toward the need to open up the rights to access and reuse research data. Given the rapid progress of the open-access movement for research data,[30] it could be expected that the next generation of researchers will be vested with greater rights to access and reuse research data compared to their predecessors. Quite naturally, their rights to access and reuse research data, comes with responsibilities as providers and users of research data. This section examines the responsibilities that must be observed by the next generation of researchers in playing their dual roles as providers and users of research data.

The Responsibilities of the Next Generation of Researchers as Data Providers

In permitting access to and reuse of research data, the data providers have the responsibility not to disclose research data that contains confidential information and not to share it with a third party.[31] From a legal perspective, nondisclosure of confidential research data implies a legal duty that further disclosure of research data to a third party will not be allowed to occur without permission or authorization from an individual or entity that discloses confidential information in the first place.[32] A nondisclosure duty could arise from contractual agreements such as a trade secret agreement, confidential agreement, or nondisclosure agreement.[33] Besides a nondisclosure duty arising from contractual agreements, it is common for the researchers, during the data collection process, to offer their promise of confidentiality that they will treat any data or information disclosed to them as confidential information. The promise of confidentiality could be given verbally or written on the consent form, interview script, survey form, recruitment letter, or brochure.[34]

---

[30] See UNESCO, 2007, Kronberg Declaration on the Future of Knowledge Acquisition and Sharing, available at <http://www.unesco.de/kronberg_declaration.html?&L=0> (accessed February 26, 2007); Organization for Economic Co-Operation and Development, 2004, Declaration on Access to Research Data From Public Funding, available at <http://www.oecd.org/document/0,2340,en_2649_34487_25998799_1_1_1_1,00.html> (accessed February 25, 2010); Council of the European Union, 2007, Council Conclusions on Scientific Information in the Digital Age: Access, Dissemination and Preservation, 1, available at <http://www.consilium.europa.eu/ueDocs/cms_Data/docs/pressData/en/intm/97236.pdf> (accessed February 28, 2010); Treaty on Access to Knowledge, 2005, available at <http://www.cptech.org/a2k/a2k_treaty_may9.pdf> (accessed March 10, 2010); International Seminar Open Access for Developing Countries, 2005, Salvador declaration on open access: The developing world perspective, available at <http://www.icml9.org/meetings/openaccess/public/documents/declaration.htm> (accessed February 26, 2010); Australian Research Information Infrastructure Committee, 2004, Open Access Statement, available at <www.caul.edu.au/caul-doc/caul2005lariic.doc> (accessed February 28, 2010); Iryna Kuchma, 2006, Berlin 4 : International Conference On Open Access, 24, available at <http://berlin4.aei.mpg.de/presentations/Kuchma_OA06.pdf> (accessed March 1, 2010); Ingegerd Rabow, 2008, Open access in Sweden: Recent development, 1 *Sciecom Info*, available at <http://www.sciecom.org/ojs/index.php/sciecominfo/article/viewFile/245/94> (accessed March 1, 2010); Hawk Jia, 2006, China Unveils Plans to Boost Scientific Data Sharing, available at <http://www.scidev.net/en/news/china-unveils-plans-to-boost-scientific-data-shari.html> (accessed March 1, 2010); Peter Suber, 2007, Pakistani Journals to be Available for Worldwide Electronic Access through Online Portal, available at <http://www.earlham.edu/~peters/fos/2007/05/pakistan-plans-oa-portal-for-all-its.html> (accessed March 1, 2010); Stevan Harnad, 2007, Brazilian Bill to Mandate OA, available at <http://www.earlham.edu/~peters/fos/2007/06/brazilian-bill-to-mandate-oa.html> (accessed March 1, 2010); see Peter Suber, 2007, Spain is Funding OA Repositories, available at <http://www.earlham.edu/~peters/fos/2007/05/spain-is-funding-oa-repositories.html> (accessed March 1, 2010).

[31] Terry Elizabeth Hendrick, 1985, Justifications for and obstacles to data sharing, in Stephen E. Fienberg, Margaret E. Martin, and Miron L. Straf, eds., *Sharing Research Data* (National Academy Press, Washington, DC) 135; Michele M. Easter, Arlene M. Davis, and Gail E. Henderson, 2004, Confidentiality: More than a linkage file and a locked drawer, 26(2) *IRB: Ethics and Human Research* 14.

[32] Howard Bauchner, 2002, Protecting research participants, 110 *Pediatrics* 402. See also Jean E. Wylie and Geraldine P. Mineau, 2003, Biomedical databases: Protecting privacy and promoting research, 21(3) *TRENDS in Biotechnology* 113.

[33] See Ian Story, 2004, Intellectual Property and Computer Software: A Battle of Competing Use and Access Visions for Countries of the South, 12; Software, available at <htt://www.answers.com/topic/computer-software> (accessed May 27, 2010).

[34] Easter, Davis, and Henderson, Confidentiality, 13.

The next generation of researchers who act as data providers also have the responsibility to protect personal or sensitive data of human subjects, usually referred to as informational privacy. The informational privacy concerns the limits on access to personal information, whereby anonymity and secrecy are branches of it.[35] Although tremendous value can be unlocked through the use of data, the fact remains that the public is concerned about the use of highly personal information.[36] For various reasons, the data subjects may not want their personal data to be revealed to a third party. This may be to avoid discrimination, personal embarrassment, or damage to their personal or professional reputation.[37] Hence, the data providers must make an important distinction between research data collected on human subjects and research data on other impersonal subjects. While research data that describes natural phenomena has a low degree of sensitivity, research data that describes humans, their activities, opinions, or behaviors may pose a high degree of sensitivity.[38]

The next generation of researchers as data providers also has the responsibility to ensure that the research data they provide to others will not prejudice national interest and security.[39] In the McKinsey report on big data, it was stated that data access can expose not only personal information and confidential corporate information but even national security and secrets.[40] Schaffer argued that advocates of open access are not proposing that institutions allow unfettered access to sensitive data that could place the security of a nation or the world at risk.[41] Although the Universal Declaration of Human Rights (1948) and the International Covenant on Civil and Political Rights (ICCPR) guarantee individuals the right to seek, receive, and impart information and ideas through any media of their choice, these rights are not without limitation.[42] In recognizing the right to impart information, the ICCPR states that the exercise of these rights carries with it special duties and responsibilities, and they are subject to certain restrictions, including for the protection of national security or of public order, as shall be provided by law.[43] Therefore, where there is a conflict between access to and reuse of research data with national security considerations, the national interest and security shall prevail.[44]

Finally, there is a responsibility for the next generation of researchers as data providers to ensure the quality and accuracy of research data that becomes the subject matter of access and reuse. The Organization for Economic Co-operation and Development principle on data quality states that data should be relevant to the purposes for which they are to be used and, to the extent necessary for those purposes, should be accurate, complete, and kept up to date.[45] While data accuracy denotes the closeness of results of observations to the true values or values accepted as being true,[46] data quality refers to the accuracy and completeness of the data and "fitness for use" for a specific dataset. It could be anticipated that a particular dataset will be unfit for use if the data and information is uncertain, inaccurate, or

---

[35] Jeantine E. Lunshof et al., 2008, From genetic privacy to open consent, 9 *Nature Reviews Genetics* 407.
[36] James Manyika et al., 2011, Big Data: The Next Frontier for Innovation, Competition, and Productivity (McKinsey Global Institute) 119.
[37] See Privacy, *Wikipedia*, available at <http://en.wikipedia.org/wiki/Privacy> (accessed June 29, 2010).
[38] Christine L. Borgman, 2005, Disciplinary differences in e-Research: an information perspective (paper presented at the International Conference on e-Social Science 2005, Manchester, UK).
[39] Stephen Hilgartner and Sherry I. Brandt-Rauf, 1994, Data access, ownership, and control: Toward empirical studies of access practices, 15 *Science Communication* 356.
[40] James Manyika et al., 2011, Big Data: The Next Frontier for Innovation, Competition, and Productivity (McKinsey Global Institute) 11.
[41] Daniel Schaffer, 2011, Free Data Has Great Value, But Challenges Remain, available at <http://www.scidev.net/en/features/free-data-has-great-value-but-challenges-remain-.html> (accessed June 28, 2011).
[42] See Universal Declaration of Human Rights 1948, Art. 19; International Covenant on Civil and Political Rights, Art. 19.
[43] See International Covenant on Civil and Political Rights, Art. 19.3
[44] Hilgartner and Brandt-Rauf, Data access, ownership, and control, 356.
[45] OECD Guidelines on the Protection of Privacy and Transborder Flows of Personal Data – Data Quality Principle.
[46] David J. Buckey, [year?] Data Accuracy and Quality, available at <http://planet.botany.uwc.ac.za/nisl/GIS/GIS_primer/page_08.htm> (accessed February 15, 2011).

erroneous.[47] Data providers have important roles to play to ensure the quality of research data.[48] The researcher who is both data creator and data provider is ultimately responsible for ensuring the quality of research data that are shared with others.[49] Being data providers, the next generation of researchers need to remind themselves that the users of research data may not know that the research data are incomplete, incompatible, or poorly documented.[50] Therefore, they need to be responsible for the quality and accuracy of research data provided by them.

The Responsibilities of the Next Generation of Researchers as Data Users
It is important for the next generation of researchers to understand that opening up the rights to access research data should not be construed as data owners having relinquished all their exclusive rights over research data. From the perspective of copyright law, opening up the rights to access and reuse research data will not in any way extinguish copyright in research data.[51] Opening up the rights to access research data does not amount to the data owner surrendering their intellectual property rights in research data. The data owners can still retain exclusive control over derivative use and commercial exploitation of research data.[52] Hence, open access to research data does not include free riding or opportunistic use, such as profit making or commercial exploitation of research data, unless specifically allowed by data owners. Since data owners still retain their rights to control use and reuse of research data, the next generation of researchers has the responsibility to ensure that they do not reuse the research data beyond the permitted rights given by data owners.[53] In the absence of explicit permission from data owners, the data users should access and reuse the research data within the scope of legitimate use provided under the fair use or fair dealing exceptions under copyright law, which permit certain uses without the need to obtain permission from data owners.

If data users need to access and reuse research data outside the scope of legitimate use provided under the fair use or fair dealing exceptions, permission must be obtained by data users from data owners.[54] In this regard, permission to access and reuse research data are subject to different licensing arrangements that require separate negotiations between data owners and data users.[55] These separate negotiations will result in different licensing rights between one licensing contract to another, which depends mostly on the

---

[47] See Jeffrey W. Seifert, 2007, CRS Report for Congress: Data Mining and Homeland Security: An Overview (Congressional Research Service) CRS-22; Earl F. Eipstein, Gary J. Hunter, and Aggrey Agumya, 1998, Liability insurance and the use of geographical information, 12(3) *Int. J. Geographical Information Science* 203; see also Aggrey Agumya and Gary J. Hunter, 1999, A risk-based approach to assessing the "fitness for use" of spatial data, 11(1) *URISA Journal* 33.

[48] Committee on Ensuring the Utility and Integrity of Research Data in [the ???]Digital Age, 2009, Ensuring the Integrity, Accessibility, and Stewardship of Research Data in the Digital Age (National Academy of Sciences, National Academy of Engineering, and Institute of Medicine, Washington, DC), 4.

[49] Ibid., 40.

[50] Jennifer L. Philips, 1999, Information liability: The possible chilling effect of tort claims against producers of geographic information systems data, 26 *Fla. St. U. L. Rev.* 749.

[51] David Shulenburger, 2003, Scholarly communications is not toxic waste: Lessons learned (prepared for the Open Access to Knowledge in the Sciences and Humanities Conference, Max Planck Society, Harnack Haus, Berlin, October) 4, available at <http://kuscholarworks.ku.edu/dspace/bitstream/1808/58/1/Scholarlypercent20Communicationspercent20ispercent20Notpercent20Toxicpercent203.pdf> (accessed October 9, 2010).

[52] See Michele Boldrin and David Levine, 2002, The case against intellectual property, 92(2) *The American Economic Review* 209; Estelle A. Fishbein, 1991, Ownership of research data, 66(3) *Academic Medicine* 129.

[53] Anne Fitzgerald, Neale Hooper, and Brian Fitzgerald, 2010, Enabling open access to public sector information with Creative Commons licenses – The Australian experience, in *Access to Public Sector Information : Law, Technology & Policy* (Sydney University Press); Linda Wang, 1999, Use of images for commercial purposes: Copyright issues under Malaysian laws, in Barbara Hoffman, ed., *Exploiting Images and Image Collections in the New Media: Gold Mine or Legal Minefield?* (Kluwer Law International, London, UK) 86.

[54] Stephanie Woods, 2008, Creative commons – A useful development in the New Zealand copyright sphere? 14 *Canterbury Law Review* 38.

[55] Wang, Use of images for commercial purposes, 86.

attitude and bargaining strength of the negotiating parties.[56] Because there could be several layers of intellectual property protection in research data, negotiations to obtain the rights to access and reuse research data could also be very cumbersome, time consuming, and excessively difficult.[57]

The next generation of researchers as data users also has the responsibility not to infringe on the moral rights of data creators. The moral rights under copyright law are meant to protect the rights of the author who creates the work.[58] Two most widely recognized moral rights are (1) the right of attribution, ensuring the data creators are recognized as the originators of their own work; and (2) the right of integrity, which allows data creators to object to mistreatment, misuse, or abuse of their works.[59] In the presence of moral right of integrity, data users are required to obtain permission from data creators or their personal representatives before they can significantly alter, modify, or distort the research data. In the absence of such permission, data users could be prohibited by data creators from making any alterations, modifications, or distortions to the research data, regardless of whether the alterations, modifications, or distortions would negatively affect or objectively improve their works.[60]

Balancing the Next Generation of Researchers' Rights to Access and Reuse Research Data with Their Responsibilities as Providers and Users of Research Data
The proponents of open access have long identified the need for a clear policy if the rights to access and reuse of research data are to be successfully implemented.[61] It was argued that, from a legal perspective, it is not possible to establish any kind of open-access system simply by default. Rather, development of an open-access system can only successfully occur through deliberate construction and active management, supported by policies and laws that facilitate the rights to access and reuse research data.[62] In a report by the American Council of Learned Societies Commission on Cyberinfrastructure for the Humanities and Social Sciences have recommendations that public and institutional policies be developed to foster openness and access not only in science and engineering research but also in humanities and social sciences research.[63]

---

[56] Academic Senate of the California State University, 2003, Intellectual Property, Fair Use, and the Unbundling of Ownership Rights (California State University) 18.
[57] Wang, Use of images for commercial purposes, 46.
[58] Cyrill P. Rigamonti, 2006, Deconstructing moral rights, 47(2) *Harvard International Law Journal* 360.
[59] See Mira T. Sundara Rajan, 2002, Moral rights and copyright harmonization: Prospects for an international moral right? (paper presented at 17th BILETA Annual Conference), available at http://www.bileta.ac.uk/02papers/sundarajan.html; Chris Armbruster, 2008, Cyberscience and the knowledge-based economy: Open access and trade publishing: From contradiction to compatibility with nonexclusive copyright licensing, (12) *International Journal of Communications Law and Policy* 17; Chris Armstrong et al., 2010, ACA2K: Comparative Review of Research Findings (Shuttleworth Foundation and University of Witwatersrand).
[60] Cyrill P. Rigamonti, Deconstructing moral rights, 49, 364.
[61] See Paul Uhlir and Peter Schroder, 2008, Open data for global science, in Brian Fitzgerald, ed., *Legal Framework for e-Research* (Sydney University Press, Sydney) 216–217; V. M. Moskovkin, 2008, Institutional policies for open access to the results of scientific research, 35(6) *Scientific and Technical Information Processing* 269; Brian Fitzgerald, Anne Fitzgerald, Professor Mark Perry, Scott Kiel-Chisholm, Erin Driscoll, Dilan Thampapillai, Jessica Coates, 2008, Creating a legal framework for copyright management of open access within the Australian academic and research sector, in Brian Fitzgerald, ed., *Legal Framework for e-Research* (Sydney University Press, Sydney) 283; Peter Arzberger et al., 2004, An international framework to promote access to data, 303 *Science* 1777; Anne Fitzgerald, Brian Fitzgerald, and Kylie Pappalardo, 2009, The future of data policy, in Tony Hey, Stewart Tansley, and Kristin Tolle, eds., *The Fourth Paradigm: Data-Intensive Scientific Discovery* (Microsoft Research, Redmond, Washington) 201; John Unsworth et al., 2006, Our Cultural Commonwealth: The Report of the American Council of Learned Societies Commission on Cyberinfrastructure for the Humanities and Social Sciences, 29, 30.
[62] See Brian Fitzgerald, Anne Fitzgerald, Professor Mark Perry, Scott Kiel-Chisholm, Erin Driscoll, Dilan Thampapillai, Jessica Coates, 2008, Creating a legal framework for copyright management of open access within the Australian academic and research sector, in Brian Fitzgerald, ed., *Legal Framework for e-Research* (Sydney University Press, Sydney) 283; Fitzgerald, Fitzgerald, and Pappalardo, Creating a legal framework, 53, 201; Anne Fitzgerald, Kylie Papalardo, and Anthony Austin, 2008, Legal implications surrounding data access, sharing and reuse, in Brian Fitzgerald, ed., *Legal Framework for e-Research* (Sydney University Press, Sydney) 161.
[63] Unsworth et al., Our Cultural Commonwealth, 29, 30.

In the Committee on Data for Science and Technology (CODATA) Berlin Conference discussion paper on international guidelines for access to research data, it was further proposed that where copyright or database law applies, the parties responsible for agreements and contracts concerning access to research data should take the relevant implications of the existing legal framework into account to allow for open access.[64] Due to the above recommendations, the next question that needs to be answered in this paper is, how should the data access and reuse policy framework that is to be developed balance the next generation of researchers' rights to access and reuse research data with their responsibilities as data providers and data users?

Balancing the Rights and Responsibilities as Open-Access Data Providers: The Responsibility Not to Disclose Research Data that Contains Confidential Information

A data access and reuse policy should balance the next generation of researchers' rights to access and reuse research data with their responsibilities as data providers by putting in place a proper mechanism to ensure the confidential data are protected while allowing access to and reuse of research data. The U.S. Panel on Data Access for Research Purposes has suggested that the confidentiality of research data be addressed by using a variety of modes for data access. The panel's suggestion involves restricting access to confidential data as well as granting unrestricted access to appropriately altered public-use data.[65] The U.S. National Research Council's Committee on National Statistics has also proposed for the technical and statistical solutions to be adopted to enable access to and reuse of data without violating a nondisclosure obligation. The statistical solutions involved data alteration, while the technical solutions involve restricted access to research data.[66] Data access and reuse policy may also require the adoption of special licensing agreements for access to confidential data in order to balance the need for data access with confidentiality protection.[67]

The Responsibility to Protect the Informational Privacy of Data Subjects

Data access and reuse policy should find a way to balance the rights to access and reuse research data with the protection of informational privacy.[68] As part of the balance between data access and data protection of informational privacy, the U.S. Privacy Protection Study Commission proposed that personal information not be disclosed to a third party in an individually identifiable form.[69] It was suggested in the report that in most cases, omitting identifiers, such as name, address, telephone number, and subject identification number, is enough to protect the anonymity of research participants.[70] Similarly the U.S. National Committee on Ensuring the Utility and Integrity of Research Data in a Digital Age also submitted in its report that for some research data, privacy protection can be addressed by removing identifiers before the sharing or public release of data.[71] Hence, a data-access and reuse policy should require data providers to remove data identifiers before sharing or public release of research data.

---

[64] Peter Schroder, 2004, Towards International Guidelines for Access to Research Data from Public Funding (Organization for Economic Co-Operation and Development) 22.
[65] Panel on Data Access for Research Purposes, 2005, Expanding Access to Research Data: Reconciling Risks and Opportunities (National Research Council) 3.
[66] National Research Council Committee on National Statistics, 2000, Improving Access to and Confidentiality of Research Data: Report of a Workshop, 29, 32.
[67] Panel on Data Access for Research Purposes, Expanding Access to Research Data, 56, 4.
[68] R. J. Bazillion, 1984, The effect of access and privacy legislation on the conduct of scholarly research in Canada, 4 *Social Science Information Studies* 7.
[69] Privacy Protection Study Commission (USA), 1997, Personal Privacy in an Information Society: The Report of the Privacy Protection Study Commission.
[70] Ibid.
[71] Committee on Ensuring the Utility and Integrity of Research Data in a Digital Age, 2009, Ensuring the Integrity, Accessibility, and Stewardship of Research Data in the Digital Age (National Academy of Sciences, National Academy of Engineering, and Institute of Medicine) 68.

Researchers who archived their research data should be requested to erase or change all names in transcripts and other material and erase any personal information that points directly to an individual.[72]

The policy should also require research data that contains sensitive and personal information to be coded, and procedures be put in place to control access.[73] Besides de-identification and coding techniques, a range of other techniques to protect informational privacy could be employed in relation to different types of research data, either qualitative or statistical. The techniques are based on reducing data specificity, distorting data, decreasing sampling sizes, and perturbing, rounding, swapping, and adding noise to the research data before release.[74] All the above technical measures could be used to protect the informational privacy while opening up the rights to access and reuse research data that contains personal information.

The Responsibility to Safeguard National Interests and Security
Data access and reuse policy should safeguard classified and secret data, where their disclosure could jeopardize national interests and security. Where national security and secrecy laws are in place to protect national interests and security, data access and reuse policy should ensure this legal duty is complied with by data providers. Therefore, there is a need to create a specific exemption in the policy, whereby the next generation of researchers is prohibited from sharing research data that is subject to national security or secrecy legislation. To balance the rights to access and reuse research data with the responsibility to safeguard national interests and security, the policy that is to be developed needs to clarify and draw a line between classified and unclassified data.[75] According to the U.S. National Academies' Committee on Ensuring the Utility and Integrity of Research Data in the Digital Age, research data pertaining to intelligence and military or terrorist activities should not be shared by researchers for security reasons.[76] Research related to nuclear, radiological, and biological threats; chemicals and explosives; human and agricultural health systems; and information technology infrastructure may also contain data that is the subject of national interests and security.[77]

The Responsibility to Ensure Data Quality and Accuracy
To ensure data quality and accuracy, data access and reuse policy should require data providers to actively supply and complete information about the research data provided.[78] If nonreviewed or nonverified research data are allowed to be shared for access and reuse, the data access and reuse policy should require data providers to warn or to notify data users of the quality and accuracy of research data. In the United States, the Office of Management and Budget's 2002 Guidelines for Ensuring and Maximizing the Quality, Objectivity, Utility, and Integrity of Information Disseminated by Federal Agencies (OMB Guidelines) require federal agencies, including the research institutions that are required by the government to disseminate publicly funded data, to adopt specific standards of quality that are appropriate for various categories of data that they disseminate.[79] To this end, the standards and quality

---

[72] Anne Sofia Fink, 2000, The role of the researcher in the qualitative research process. A potential barrier to archiving qualitative data, 1(3) *Forum: Qualitative Social Research*, available at <http://www.qualitative-research.net/index.php/fqs/article/viewArticle/1021/2201> (accessed July 2, 2011).
[73] Beatrice Godard et al., 2003, Data storage and DNA banking for biomedical research: Informed consent, confidentiality, quality issues, ownership, return of benefits. A professional perspective, 11(Suppl 2) *European Journal of Human Genetics* S91.
[74] Mark Elliot, Kingsley Purdam, and Duncan Smith, 2005, Confidential data access using grid computing: An outline of the issues and possible solutions (paper presented at the International Conference on e-Social Science 2005, Manchester Conference Center, June 22–24).
[75] Committee on Ensuring the Utility and Integrity of Research Data in a Digital Age, Ensuring the Integrity, Accessibility, and Stewardship of Research Data, 62.
[76] Ibid., 68.
[77] Ibid.
[78] Earl F. Eipstein, Gary J. Hunter, and Aggrey Agumya, 1998, Liability insurance and the use of geographical information, 12(3) *Int. J. Geographical Information Science* 210.
[79] Guidelines for Ensuring and Maximizing the Quality, Objectivity, Utility and Integrity of Information Disseminated by Federal Agencies 2002, Guidelines III, para.1.

to ensure data quality provided under the OMB Guidelines could be used by data access and reuse policy for the purpose of imposing a standard of care on data providers to ensure data quality and accuracy.

Balancing the Rights and Responsibilities as Open-Access Data Users: The Responsibility Not to Use or Reuse Open-Access Research Data Beyond The Permitted Rights

In most countries, the fair use or fair dealing exceptions that dispense the need to obtain the data owner's permission is limited to access and reuse of research data for private and educational purposes only. While the next generation of researchers require the right to access and reuse research data to be opened, the restrictive scope of legitimate use under fair use or fair dealing exceptions under copyright law could prevent data users from exploiting the full value and potential of research data.[80] For the purpose of opening up the rights to access and reuse research data, it was argued by Graham Greenleaf that the rights to access and reuse the research data given by data owners should be more extensive and beyond fair use or fair dealing exceptions offered by the copyright law.[81] For Uhlir and Schroder, open access in the context of public research data should be interpreted as access on equal terms for the international research community, as well as industry, with the fewest restrictions on reuse.[82] This could well mean that the next generation of researchers should be given broad rights to access and reuse research data, including for profit, industrial, commercial, or nonacademic research purposes.

The Responsibility to Obtain Permission to Access and Reuse Research Data

One of the critical aspects in opening up the rights to access and reuse research data are to ensure that the rights are not only given to the first user but that they remain freely accessible and usable by downstream users.[83] In granting the rights to access and reuse data, data owners can choose from a wide range of licensing conditions, from the widest possible license to the narrowest form of licensing.[84] If data users have to enter into licensing or assignment contracts with data owners each time they require access to and reuse of research data, the progress of research will be retarded and transaction cost for data access will be high.[85] Even when the rights to access and reuse is granted by data owners by way of licensing, the copyright licensing mechanism is said to be time-consuming, and not well suited to the digital environment.

To balance the need to access and reuse research data with the responsibilities to obtain licensing rights, a data access and reuse policy should require data owners to adopt the simplest form of licensing scheme that accelerates and produces the optimum rights to access and reuse research data.[86] To this end, the policy should require data owners to give permission in advance to data users to access and reuse the research data. This advance permission should be given in a ready-made template by using standard licensing schemes such as Creative Commons licenses, Science Commons licenses, or GNU General Public License. In this regard, the better approach for promoting open access to data is to apply for a

---

[80] Nicole Ebber, 2008, Creative Commons licenses: New ways of granting and utilizing access to information (paper presented at the 16th BOBCATSSS Symposium 2008 – Providing Access to Information for Everyone, Zadar, Croatia, January 28–30).
[81] Graham Greenleaf, 2008, Unlocking IP to Stimulate Australian Innovation: An Issues Paper (University of New South Wales). See also P. Arzberger et al., 2004, Promoting access to public research data for scientific, economic, and social development, 3 *Data Science Journal* 146; Kai Ekholm, Access to Our Digital Heritage
[82] Paul Uhlir and Peter Schroder, 2008, Open data for global science, in Brian Fitzgerald, ed., *Legal Framework for e-Research: Realising the Potential* (Sydney University Press Sydney) 209.
[83] Karl-Nikolaus Peifer, 2008, Open access and (German) copyright, in *Open Access: Opportunities and Challenges – A Handbook* (UNESCO) 50.
[84] Woods, Creative commons, 45.
[85] Gideon Emcee Christian, 2009, Building a Sustainable Framework for Open Access to Research Data Through Information and Communication Technologies (International Development Research Center (IDRC) Canada) 22.
[86] Ibid., 75.

Creative Commons Attribution (CC-BY) license to the data. This allows the data to be widely shared and used, but also preserves the creator's right to attribution.[87]

The Responsibility Not to Infringe on Data Creators' Moral Rights
To open up the rights to access and reuse research data for the next generation of researchers, the moral right of integrity should be construed in such a way as to ensure that the personal interests of authors do not interfere with the legitimate self-expression of future authors.[88] Data access and reuse policy should provide guidelines about what sort of acts are considered as infringing on data creators' moral rights. To this end, there should be clear guidelines about the reasonable circumstances that allow alteration or modification of the research data without being construed as an infringement of data creators' moral rights of integrity. There should also be exceptions to data creators' moral rights of integrity for certain categories of copyright works as found in the Australian Copyright Amendment (Moral Rights) Act 2000 (Commonwealth of Australia),[89] the U.K. Copyright, Designs and Patents Act (CDPA) 1988,[90] and the U.S. Visual Artists Rights Act of 1990.[91] Further, in the U.K. CDPA,[92] the moral right to integrity does not exist for copyright work produced under work for hire, that is, arising from employment or commissioned work. Applying the U.K. approach, this paper argues that, where the research data are publicly funded or produced by data creators under work for hire, they should be required to waive their moral right to integrity, which enables the research data to be transformed into beneficial use by other researchers.

Conclusion
This presentation argues that to avoid the next generation of researchers suffering the same disadvantages as their predecessors, it is deemed necessary to open up the rights to access and reuse research data. Premised upon this argument, this paper further argues that a policy framework needs to be developed to balance the rights and responsibilities of the next generation of researchers as providers and users of research data. Therefore, apart from establishing the need to open up the rights to access and reuse research data for the next generation of researchers, this paper has also put forward several recommendations for balancing the rights to access and reuse research data with the responsibilities outlined above. Most of these recommendations would be best carried out by the research funders and the research institutions of each country.[93]

Despite focusing on the rights and responsibilities of the next generation of researchers, it is clear that these rights and responsibilities are applicable regardless of the generation with which the researchers may identify. However, as it is predicted that data access and reuse will be opened at a far larger scale in the near future, awareness towards the responsibilities pertaining to access to and reuse of research data, both as providers and users, is highly expected from the next generation of researchers. It is believed that through a proper policy framework that balances the rights to access and reuse with the responsibilities as

---

[87] Anne Fitzgerald, 2009, Sharing Environmental Data: The Role of an Information Policy Framework and Copyright Licensing, available at <http://www.eresearch.edu.au/docs/2009/era09_submission_122.pdf> (accessed June 20, 2011);
[88] Nicolas Suzor, 2006, Transformative Use of Copyright Material (Master's thesis, Queensland University of Technology).
[89] Under the Australian Copyright Act, the author's right of integrity does not apply to sound recording, but is applicable to literary, dramatic, musical, artistic, and cinematography works. See Copyright Amendment (Moral Rights) Act 2000 (Commonwealth of Australia), Sec. 195AJ, 195AK, 195AL – Division 4—Right of integrity of authorship of a work.
[90] The moral right of integrity under the U.K. CDPA does not apply to computer programs or to any computer-generated works. See Copyright, Designs and Patents Act 1988, c. 48 (Eng.), Sec. 81(2) – Exceptions to right.
[91] The author's moral right of integrity in the United States does not apply to all authors, but is only applicable to certain authors of the specified group of works. See U.S. Code Title 17 Copyright Act of 1976, s. 106A – Rights of certain authors to attribution and integrity. See also Visual Artists Rights Act of 1990 (US), Sec. 101(1), 101(2) – Work of Visual Art Defined.
[92] Copyright, Designs and Patents Act 1988, c. 48 (Eng.), Sec. 82(1)(a), (b) – Qualification of right in certain cases.
[93] Uhlir and Schroder Open data for global science, 188.

data providers and data users, the next-generation researcher will be able to enjoy greater, yet well-governed, rights to access and reuse research data.

REFERENCES

Academic Senate of the California State University, Intellectual Property, Fair Use, and the Unbundling of Ownership Rights (California State University, 2003) 18.

Agumya, Aggrey, and Gary J. Hunter, A risk-based approach to assessing the "'fitness for use" of spatial data (1999) 11(1) *URISA Journal* 33.

Andanda, P. A., Human-tissue-related inventions: Ownership and intellectual property rights in international collaborative research in developing countries (2006) 34 *J Med Ethics* 172.

Armbruster, Chris, Cyberscience and the knowledge-based economy: Open access and trade publishing: From contradiction to compatibility with nonexclusive copyright licensing (2008) (12) *International Journal of Communications Law and Policy* 17.

Armstrong, Chris, et al., ACA2K: Comparative Review of Research Findings (Shuttleworth Foundation and University of Witwatersrand, 2010).

Arzberger, P., et al., Promoting access to public research data for scientific, economic, and social development (2004) 3 *Data Science Journal* 146. Kai Ekholm, 'Access to Our Digital Heritage'

Arzberger, Peter, et al., An international framework to promote access to data (2004) 303 *Science* 1777.

Atkins, Daniel E., et al., Revolutionizing Science and Engineering Through Cyberinfrastructure: Report of the National Science Foundation Blue-Ribbon Advisory Panel on Cyberinfrastructure (National Science Foundation, 2003) 17, 28.
Australian Research Information Infrastructure Committee, Open Access Statement (2004), available at <www.caul.edu.au/caul-doc/caul2005 1ariic.doc> (accessed February 28, 2010).

Bauchner, Howard, Protecting research participants (2002) 110 *Pediatrics* 402. See also Jean E. Wylie and Geraldine P. Mineau, Biomedical databases: Protecting privacy and promoting research (2003) 21(3) *TRENDS in Biotechnology* 113.

Bazillion, R. J., The effect of access and privacy legislation on the conduct of scholarly research in Canada (1984) 4 *Social Science Information Studies* 7.

Benkler, Yochai, The university in the networked economy and society (2008) (November/December) *Educause* 65.

Berman, Francine, and Henry E. Brady, Workshop on Cyberinfrastructure for the Social and Behavioral Sciences: Final Report (National Science Foundation, 2005) 32, 57.

Bicknell, David, Collaboration drives innovation (2009) (Apr 28–May 4) *Computer Weekly* 14.

Blumberg, Peter D., From "publish or perish" to "profit or perish": Revenues from university technology transfer and the s. 501(c)(3) tax exemption (1996) 145(1) *The University of Pennsylvania Law Review* 91.

Boldrin, Michele, and David Levine, The case against intellectual property (2002) 92(2) *The American Economic Review* 209.

Borgman, Christine L., Disciplinary Differences in e-Research: An Information Perspective (paper presented at the International Conference on e-Social Science 2005, Manchester, UK, 2005).

Boyle, James, Foreword: The opposite of property? (2003) 66(1&2) *Law & Contemporary Problems* 13.

Buckey, David J., Data Accuracy and Quality, available at <http://planet.botany.uwc.ac.za/nisl/GIS/GIS_primer/page_08.htm> (accessed February 15, 2011).

Christian, Gideon Emcee, Building a Sustainable Framework for Open Access to Research Data Through Information and Communication Technologies (International Development Research Center [IDRC] Canada, 2009) 22.

Christie, Andrew F., et al., Analysis of the Legal Framework for Patent Ownership in Publicly Funded Research Institutions (Commonwealth of Australia Department of Education, Science and Training, 2003) 48.

Committee on Ensuring the Utility and Integrity of Research Data in a Digital Age, Ensuring the Integrity, Accessibility, and Stewardship of Research Data in the Digital Age (National Academy of Sciences, National Academy of Engineering, and Institute of Medicine, 2009) 4, 40, 68.

Committee on National Statistics, Issues and recommendations, in Stephen E. Fienberg, Margaret E. Martin, and Miron L. Straf (eds.), *Sharing Research Data* (National Academy Press, Washington DC, 1985) 24, 25.

Council of the European Union, Council Conclusions on Scientific Information in the Digital Age: Access, Dissemination and Preservation (2007) 1, available at <http://www.consilium.europa.eu/ueDocs/cms_Data/docs/pressData/en/intm/97236.pdf> (accessed February 28, 2010).

Czarnitzki, Dirk, Wolfgang Glanzel, and Katrin Hussinger, Heterogeneity of patenting activity and its implications for scientific research (2009) 38 *Research Policy* 33.

David, Paul A., Can "open science" be protected from the evolving regime of IPR protections? (2004) 129 (March) *Journal of Institutional and Theoretical Economics* 3.

Easter, Michele M., Arlene M. Davis, and Gail E. Henderson, Confidentiality: More than a linkage file and a locked drawer (2004) 26(2) *IRB: Ethics and Human Research* 14.

Ebber, Nicole, Creative Commons Licenses: New Ways of Granting and Utilizing Access to Information (paper presented at the 16th BOBCATSSS Symposium 2008 – Providing Access to Information for Everyone, Zadar, Croatia, January 28–30).

Eipstein, Earl F., Gary J. Hunter, and Aggrey Agumya, Liability insurance and the use of geographical information (1998) 12(3) *Int. J. Geographical Information Science* 203, 210.

Elliot, Mark, Kingsley Purdam, and Duncan Smith, Confidential Data Access Using Grid Computing: An Outline of the Issues and Possible Solutions (paper presented at the International Conference on e-Social Science 2005, Manchester Conference Center, June 22–24, 2005).

Fink, Anne Sofia, The role of the researcher in the qualitative research process. A potential barrier to archiving qualitative data (2000) 1(3) *Forum: Qualitative Social Research*, available at <http://www.qualitative-research.net/index.php/fqs/article/viewArticle/1021/2201> (accessed July 2, 2011).

Fishbein, Estelle A., Ownership of research data (1991) 66(3) *Academic Medicine* 129.

Fitzgerald, Anne, Sharing Environmental Data: The Role of an Information Policy Framework and Copyright Licensing (2009), available at <http://www.eresearch.edu.au/docs/2009/era09_submission_122.pdf> (accessed June 20, 2011).

Fitzgerald, Anne, Brian Fitzgerald, and Kylie Pappalardo, The future of data policy, in Tony Hey, Stewart Tansley, and Kristin Tolle (eds.), *The Fourth Paradigm: Data-Intensive Scientific Discovery* (Microsoft Research, Redmond, Washington, 2009) 201.

Fitzgerald, Anne, Neale Hooper, and Brian Fitzgerald, Enabling open access to public sector information with Creative Commons licenses – The Australian experience, in *Access to Public Sector Information : Law, Technology and Policy* (Sydney University Press, 2010).

Fitzgerald, Anne, and Kylie Pappalardo, Building the Infrastructure for Data Access and Reuse in Collaborative Research: An Analysis of the Legal Context (2007).

Fitzgerald, Anne, Kylie Papalardo, and Anthony Austin, Legal implications surrounding data access, sharing and reuse, in Brian Fitzgerald (ed.), *Legal Framework for e-Research* (Sydney University Press, Sydney, 2008) 161.

Fitzgerald, Brian, Anne Fitzgerald, Professor Mark Perry, Scott Kiel-Chisholm, Erin Driscoll, Dilan Thampapillai, Jessica Coates, Creating a legal framework for copyright management of open access within the Australian academic and research sector, in Brian Fitzgerald (ed.), *Legal Framework for e-Research* (Sydney University Press, Sydney, 2008) 283.

Fry, Jenny, Ralph Schroeder, and Matthijs des Besten, Open science in e-Science: Contingency or policy? (2009) 65(1) *Journal of Documentation* 7.

Glinos, Kostas, Report of the High Level Expert Group on Scientific Data: Riding the Wave: How Europe Can Gain From the Rising Tide of Scientific Data – A Vision for 2030 (European Commission, 2010).

Godard, Beatrice, et al., Data storage and DNA banking for biomedical research: Informed consent, confidentiality, quality issues, ownership, return of benefits. A professional perspective, (2003) 11(Suppl 2) *European Journal of Human Genetics* S91.

Greenleaf, Graham, Unlocking IP to Stimulate Australian Innovation: An Issues Paper (University of New South Wales, 2008).

Guidelines for Ensuring and Maximizing the Quality, Objectivity, Utility and Integrity of Information Disseminated by Federal Agencies 2002, Guidelines III, para.1.

Harnad, Stevan, Brazilian Bill to Mandate OA (2007), available at <http://www.earlham.edu/~peters/fos/2007/06/brazilian-bill-to-mandate-oa.html> (accessed March 1, 2010).

Hendrick, Terry Elizabeth, Justifications for and obstacles to data sharing, in Stephen E. Fienberg, Margaret E. Martin, and Miron L. Straf (eds.), *Sharing Research Data* (National Academy Press, Washington DC, 1985) 135.

Hess, Charlotte, and Elinor Ostrom, Ideas, artifacts, and facilities: Information as a common-pool resource (2003) 66 *Law and Contemporary Problems* 112.

Hey, Tony, and Anne E. Trefethen, Cyberinfrastructure for e-Science (2005) 308 *Science Mag* 818.

Hey, Tony, Stewart Tansley, and Kristin Tolle (eds.), *The Fourth Paradigm: Data-Intensive Scientific Discovery* (Microsoft Research, Redmond, Washington, 2009).

Hilgartner, Stephen, and Sherry I. Brandt-Rauf, Data access, ownership, and control: Toward empirical studies of access practices' (1994) 15 *Science Communication* 356.

International Seminar on Open Access for Developing Countries, Salvador Declaration on Open Access: The Developing World Perspective (2005), available at <http://www.icml9.org/meetings/openaccess/public/documents/declaration.htm> (accessed February 26, 2010).

Jia, Hawk, China Unveils Plans to Boost Scientific Data Sharing (2006), available at <http://www.scidev.net/en/news/china-unveils-plans-to-boost-scientific-data-shari.html> (accessed March 1, 2010).

Kirsop, Barbara, and Leslie Chan, Transforming access to research literature for developing countries (2005) 31(4) *Serials Review* 247.
Kuchma, Iryna, Berlin 4: International Conference on Open Access (2006) 24, available at <http://berlin4.aei.mpg.de/presentations/Kuchma_OA06.pdf> (accessed March 1, 2010).
Lunshof, Jeantine E., et al., From genetic privacy to open consent (2008) 9 *Nature Reviews Genetics* 407.

Manyika, James, et al., Big Data: The Next Frontier for Innovation, Competition, and Productivity (McKinsey Global Institute, 2011) 11, 119.

Moskovkin, V. M., Institutional policies for open access to the results of scientific research (2008) 35(6) *Scientific and Technical Information Processing* 269.

National Research Council Committee on National Statistics, Improving Access to and Confidentiality of Research Data: Report of a Workshop (??National Academies Press, Washington, DC, 2000) 29, 32.

Nelson, Michael L., Data-driven science: A new paradigm? (2009) 44(4) *Educause* 6.
OECD Guidelines on the Protection of Privacy and Transborder Flows of Personal Data – Data Quality Principle.

Open Access to Knowledge (OAK) Law Project: Legal Protocols for Copyright Management: Facilitating Open Access to Research at the National and International Levels, available at <http://www.oaklaw.qut.edu.au/files/Data_Report_final_web.pdf> (accessed January 11, 2010) 58.

Organization for Economic Co-Operation and Development, Declaration on Access to Research Data From Public Funding (2004), available at <http://www.oecd.org/document/0,2340,en_2649_34487_25998799_1_1_1_1,00.html> (accessed February 25, 2010).

Panel on Data Access for Research Purposes, Expanding Access to Research Data: Reconciling Risks and Opportunities (National Research Council, 2005) 3, 4.

Peifer, Karl-Nikolaus, Open access and (German) copyright, in *Open Access: Opportunities and Challenges – A Handbook* (UNESCO, 2008) 50.

Philips, Jennifer L., Information liability: The possible chilling effect of tort claims against producers of geographic information systems data (1999) 26 *Fla. St. U. L. Rev.* 749.

Privacy Protection Study Commission (USA), Personal Privacy in an Information Society: The Report of the Privacy Protection Study Commission (1977).

Privacy, *Wikipedia*, available at <http://en.wikipedia.org/wiki/Privacy> (accessed June 29, 2010).

Rabow, Ingegerd, Open access in Sweden: Recent development (2008) (1) *Sciecom Info*, available at <http://www.sciecom.org/ojs/index.php/sciecominfo/article/viewFile/245/94> (accessed March 1, 2010).

Rajan, Mira T. Sundara, Moral Rights and Copyright Harmonization: Prospects for an International Moral Right? (2002) (paper presented at the 17th BILETA Annual Conference), available at http://www.bileta.ac.uk/02papers/sundarajan.html.

Rambo, Neil, E-science and biomedical libraries (2009) 97(3) *J Med Libr Assoc* 160.

Reichman, Jerome H., and Tracy Lewis, Using liability rules to stimulate local innovation in developing countries: Application to traditional knowledge, in Keith Eugene Maskus and Jerome H. Reichman (eds.), *International Public Goods and Transfer of Technology Under a Globalized Intellectual Property Regime* (Cambridge University Press, 2005) 340.

Research Councils UK, Open Access to Research Outputs: Final Report to RCUK (2008) 21, available at <http://www.rcuk.ac.uk/cmsweb/downloads/rcuk/news/oareport.pdf> (accessed March 11, 2010).

Rigamonti, Cyrill P., Deconstructing moral rights (2006) 47(2) *Harvard International Law Journal* 360.

Rose, Carol M., Romans, roads, and romantic creators: Traditions of public property in the information age (2003) 66 *Law & Contemporary Problems* 90.

Royal Society (United Kingdom), Keeping Science Open: The Effects of Intellectual Property Policy on the Conduct of Science (2003) 21, available at <http://royalsociety.org/Report_WF.aspx?pageid=9842&terms=Keeping+science+openpercent3A+the+effects+of+intellectual+property+policy+on+the+conduct+of+science> (accessed February 26, 2010)

Salokannel, Marjut, Global Public Goods and Private Rights: Scientific Research and Intellectual Property Rights (2003) 11, available at <http://www.iprinfo.com/tiedostot/5icFWowu.pdf> (accessed October 12, 2010).

Sampat, Bhaven N., Patenting and US academic research in the 20th century: The world before and after Bayh-Dole (2006) 35 *Research Policy* 784.

Samuelson, Pamela, Mapping the digital public domain: Threats and opportunities (2003) 66(1 and 2) *Law & Contemporary Problems* 170.

Schaffer, Daniel, Free Data has Great Value, But Challenges Remain (2011), available at <http://www.scidev.net/en/features/free-data-has-great-value-but-challenges-remain-.html> (accessed June 28, 2011).

Schroder, Peter, Towards International Guidelines for Access to Research Data From Public Funding (Organization for Economic Co-Operation and Development, 2004).

Science and Technology Committee, Science and Technology – Tenth Report (U.K. House of Commons, Science and Technology Committee, 2004).

Shulenburger, David, Scholarly communications is not toxic waste: Lessons learned (prepared for the Open Access to Knowledge in the Sciences and Humanities Conference, Max Planck Society, Harnack Haus, Berlin, October 2003) 4, available at <http://kuscholarworks.ku.edu/dspace/bitstream/1808/58/1/Scholarlypercent20Communicationspercent20ispercent20Notpercent20Toxicpercent203.pdf> (accessed October 9, 2010).

Schwartz, Charles A., Reassessing prospects for the open access movement (2005) (November) *College & Research Libraries* 491.

Seifert, Jeffrey W., CRS Report for Congress: Data Mining and Homeland Security: An Overview (Congressional Research Service, 2007) CRS-22.

Story, Ian, Intellectual property and computer software: A battle of competing use and access visions for countries of the south (2004) 12; *Software*, available at <htt://www.answers.com/topic/computer-software> (accessed May 27, 2010).

Suber, Peter, Pakistani Journals to be Available for Worldwide Electronic Access through Online Portal (2007), available at <http://www.earlham.edu/~peters/fos/2007/05/pakistan-plans-oa-portal-for-all-its.html> (accessed March 1, 2010).

——, Spain is Funding OA Repositories (2007), available at <http://www.earlham.edu/~peters/fos/2007/05/spain-is-funding-oa-repositories.html> (accessed March 1, 2010).

Suzor, Nicolas, Transformative Use of Copyright Material (Master's thesis, Queensland University of Technology, 2006).

Swan, Alma, Open access and the progress of science (2007) 95(May-June) *American Scientist* 200.

Thursby, Jerry G., Richard Jensen, and Marie G. Thursby, Objectives, characteristics and outcomes of university licensing: A survey of major U.S. universities (2001) 26 *Journal of Technology Transfer* 59.

Treaty on Access to Knowledge (2005), available at <http://www.cptech.org/a2k/a2k_treaty_may9.pdf> (accessed March 10, 2010).

Trosow, Samuel E., Copyright Protection for Federally Funded Research: Necessary Incentive or Double Subsidy? (2003) 18, available at <http://publish.uwo.ca/-strosow/Sabo_Bill_Paper.pdf> (accessed September 12, 2010).

Uhlir, Paul, and Peter Schroder, Open data for global science, in Brian Fitzgerald (ed.), *Legal Framework for e-Research* (Sydney University Press, Sydney, 2008) 188, 209, 216–217.

UNESCO, Kronberg Declaration on the Future of Knowledge Acquisition and Sharing (2007), available at <http://www.unesco.de/kronberg_declaration.html?&L=0> (accessed February 26, 2007).

Unsworth, John, et al., Our Cultural Commonwealth: The Report of the American Council of Learned Societies Commission on Cyberinfrastructure for the Humanities and Social Sciences (2006) 29, 30.

Wang, Linda, Use of images for commercial purposes: Copyright issues under Malaysian laws, in Barbara Hoffman (ed.), *Exploiting Images and Image Collections in the New Media: Gold Mine or Legal Minefield?* (Kluwer Law International, London, 1999) 86.

Woods, Stephanie, Creative Commons – A useful development in the New Zealand copyright sphere? (2008) 14 *Canterbury Law Review* 38.

33. DISCUSSION BY THE WORKSHOP PARTICIPANTS

PARTICIPANT: My question is addressed to Mr. Schaffer, but I invite everybody on the panel to comment and express their point of view. About a year ago, I read a publication of science metrics, which measures the level of scientific activity in the world, and there were comparisons of different countries. China and Turkey, which you mentioned, show a really high level of scientific activity, but what struck me more was Iran and the huge spike in scientific activities there, especially related to fundamental and natural sciences. Basically, I invite you to comment on this trend related to data-sharing policies and the global nature of data-sharing policies, open data to scientific and research information, and maybe facilitating a dialogue. I am not sure if there are any partnerships between your organizations and Islamic countries apart from Egypt, which was mentioned. I would appreciate your views.

MR. SCHAFFER: I will speak on behalf of the Academy of Sciences for the Developing World (TWAS) and others can join. The roots of TWAS and its identity lie in its efforts to build scientific capacity. Much of the work of TWAS over the course of its existence has been devoted to investments and partnerships for capacity building. Most of that has been in training and research. As the capacity of various countries improved, we have moved increasingly toward South-South cooperation. In some areas, we see some lead countries, for example, China in Asia, South Africa in Africa, and Brazil in South America. Our hope is that we can develop South-South cooperation networks, not just in publication output but also in projects that have impact on the ground, and do that by allowing the advancing countries in the South to take the lead in South-South partnerships. One of the critical questions or issues that TWAS is examining is whether countries like China, Brazil, and India will find greater reason to ally with the United States and Europe than with other countries in the developing world. It is that kind of dynamic that is now in play for scientific projects.

DR. RUMBLE: I would like to add that if you look at the demographics of the northern African countries all the way into Pakistan, the percentage of people who are under 25 and even under 15 is really startling compared to societies such as ours. I think organizations like TWAS and the United Nations Educational, Scientific and Cultural Organization (UNESCO) have a unique opportunity to affect the young generation through science. There is tremendous opportunity for this generation to get turned on to science and see what it can do for them individually, as well as for the institutions that they cherish and their countries. I certainly hope that catches on.

MR. SCHAFFER: There are limits to what organizations like TWAS and UNESCO can do. Some of the discussion this morning was about politics. There are political issues that go beyond science education and training. Still, we can argue that, in some ways, the investments in science can set the stage. Education can set the stage for political change.

DR. GRIFFIN: I sponsored three workshops on building a digital library in the Middle East region. One was in 2006 at the New Alexandria Library; one in Rabat, Morocco, in 2007; and another one here in Washington, D.C., on January 24 and 25, 2011. We need to keep in mind when we are talking about individual countries in different parts of the world that there are shared cultural identities that go across national and political boundaries. The general approaches, feelings, and values regarding the way in which they receive the world and the way in which they receive the world through the scientific method is somewhat different in each of these cultures. Therefore, more communication mechanisms need to be established as an intermediate step to what we might be trying to get in the long term.

PARTICIPANT: I wanted to get to that issue of incentives that were raised in a couple of talks. It is an important issue, especially because many organizations have been talking about scientific publications and open-access publications. Some of you may remember that maybe a decade or so ago, people were encouraged to do interdisciplinary work, but many people who published in interdisciplinary journals did not get tenure and they did not get promoted. People did not like it. I have not heard too much discussion about how to change the incentive structures so that we do not have these problems with young people.

DR. GUSTAFSSON: Traditionally we have had the habit of saying that when writing a paper or proposal, you should be able to put your work into a more general framework. It is still disciplinary and is still only related to science itself, but it has a context. I think we can say that any scientist who is worth supporting should be able to put his or her work into a more general framework related to the world today. It is hard to make it an absolute condition, but I think it could be a standard approach. I think research-funding agencies could put more emphasis on these things.

PARTICIPANT: I would like to thank Bengt Gustafsson for his opening remarks emphasizing the importance of free travel, communication, and collaboration as an element of international science. The American Physical Society (APS) has been very active in this regard. The APS has an Office of International Affairs and runs the Forum on International Physics, which follows up on international issues. I would like to share with you one experience I had as chair of the forum in 2007. You may remember that in 2007 the American Chemical Society (ACS) decided on its own that it had to expel all Iranian members. They misinterpreted some of the sanctions rules, which in fact specifically allow the exchange of open scientific literature among members around the world without being subject to sanctions, but nevertheless they expelled them. The APS, especially the Forum on International Physics, took major action and wrote many letters protesting this. As a result, the ACS backed down and reinstated the Iranians.

The interesting thing is that shortly after this event, the University and College Union of the United Kingdom decided to institute a boycott against Israeli academics. We took action against that as well. I took the opportunity to inform many colleagues and friends I had in Iran and elsewhere in the Middle East of what was going on. I said if you have any comments or thoughts about this, why not write to the University and College Union and tell them what you think. I will read to you two messages that came from friends in the Middle East. The first one was from Egypt and said, "I regard the collective punishment of Israeli scientists to be unfair. In spite of the collective punishment practice by the Israeli government against Palestinians, scientists should not be punished just because of their nationality." The second note was even more dramatic, from a very prominent Iranian scientist, president of the Iranian Physical Society and deputy minister of science in Iran, who said, "Let me express my sincere opposition to the boycott of Israeli academics that is being considered by the University and College Union. As a scientist living in the Middle East, I appreciate the move of your organization to express its unhappiness about the restrictions being made by Israeli forces on Palestinian students and academics. However, the decision made by your organization is violating the same principles that you are trying to defend. It is hard to accept that the Israeli academics are proponents of such restrictions." These examples show that as we engage in collaborations and communications with our colleagues abroad, we become aware of incidents like this (human rights violations, academic freedom violations), and we can take action using our connections abroad to influence it.

DR. GUSTAFSSON: I certainly agree. I think you will also agree that this experience shows that this kind of activity is necessary, but persistence is also needed. In other words, a continuous effort must be made here. There is another point I also want to stress. We can hear other people claiming that this defense of universality is just in the scientific self-interest. I think that is not right, because we all believe that the efforts made would give more reward back to society.

DR. KAHN: It is not just about data. Data are the progeny of scientific activity, and it is everything that goes into that. Something that strikes me as obvious that has hardly been spoken about is the continuing divide between the natural sciences and engineering and the social sciences, which is as important for data and progress of sciences. That is one observation. The other one, which has not really come up in the discussions very much, is that when we talk about scientific collaboration in the global context, it tends to be North-South rather than South-South. It is a continuing tragedy and very easy to establish. You simply go to a Web of Science[94] and do your analysis and you will see. The point I am making is that the South tends to collaborate first with the North rather than with the South. Changing that is certainly one of the most important challenges we have ahead of us.

DR. INYANG: Yes. I think this is one of the toughest things to change in Africa. This is precisely why I mentioned it in my presentation. There will not be movement away from that circumstance and there will not be South-South collaboration unless there is a continental-level institution to promote and facilitate South-South in Africa. For example, if we have solicitations that say we would like a university from South Africa to collaborate with a university in Senegal, it is very difficult to get that to happen, because there are many constraints like language, university systems, and so on. Also, most of the universities do not have the structure that is common in the United States. This is a big issue, and the proposed African science foundation would help in this area. Moreover, most international agencies do not have a formal research solicitation program that is specifically targeting African scientists. UNESCO is trying. Sometimes they have an African-wide research solicitation, but that pales in comparison to the enormity of issues and problems that Africa needs to confront. The aid given by the European Union and the U.S. Agency for International Development and others should at least have considered the creation of such an entity, because unless there is an entity like that, most of the people who have Ph.D.'s in Africa do not really have a place to go. Some of them migrate into government agencies, where the requirements that they use their intellect to do things are secondary to political patronage.

PARTICIPANT: We just heard a series of approaches from different sectors of the scientific community. What I would like everybody to comment on is whether they feel that the funding and the commitment of the people who could provide funding is there, or whether these are wonderful programs, wonderful approaches, and wonderful ideas, but we are so lacking in funding that progress is going to be significantly hindered.

MR. SCHAFFER: Actually, I did not mention in the presentation that TWAS has been very fortunate, because it has had base funding provided by the Italian government with very little oversight of direction. Italy basically has given TWAS core funding since its inception and has allowed it the freedom to do what it thought needed to be done to accomplish or advance its goals. Being freed from looking for core funding has enabled TWAS to take risks, explore different avenues and different opportunities knowing that its core funding and its core staff would be financially protected. I think that has made a huge difference, and it is an example in a sense of North-South cooperation at a reasonable level of effectiveness. A major part of the reason, I think, has been the funding, and then the freedom to move on the part of the organization that has received the funding.

DR. GRIFFIN: I can only speak from the experience of the digital libraries initiatives. We found that some of the most valuable long-term outcomes were those associated with process rather than simply individual basic research grants. For example, we fund working groups that develop positions on issues relevant to information management. I think that the next development that is going to have a significant effect on the way in which we think about scientific endeavor are new document-publication models. When the publication model changes from papers to actively formatted documents, when we have links to datasets and other publications, when other people can use the same dataset and see if they can replicate our experiments, we are going to have an altogether different perspective on how we do science collectively.

DR. INYANG: I think one of the things that can be done is to create continental support systems, because

---

[94] Available at http://thomsonreuters.com/products_services/science/science_products/a-z/web_of_science/

while South Africa, Tunisia, Algeria, Egypt, and perhaps Nigeria can have the support systems to help themselves, there are many other countries that are just too poor to support themselves. They do not necessarily see investment in science as something important.

APPENDIXES

Appendix A

Meeting Agenda

**Board on International Scientific Organizations
and the
U.S. Committee on Data for Science and Technology
Board on Research Data and Information
National Academy of Sciences
in consultation with the
Committee on Freedom and Responsibility in the Conduct of Science
International Council for Science**

Room 100
National Academy of Sciences
500 Fifth Street NW
Washington, DC

April 18–19, 2011

**Day One:
Session One: Setting the Stage
Session Chair: <u>Farouk El-Baz</u>, Boston University**

8:45	*Welcoming remarks*	Charles Vest, President, National Academy of Engineering
9:00	Background and purpose of the symposium: A historical perspective	Farouk El-Baz, Symposium co-chair
9:20	Keynote presentation: Why is international scientific data sharing important?	Atta-ur-Rahman, UNESCO Science Laureate
10:00	*Coffee break*	

**Session Two: Status of Access to Scientific Data
Session Chair: <u>Roberta Balstad</u>, Columbia University**

10:30	Overview of scientific data policies	Roberta Balstad, Columbia University, United States
	Examples of scientific data-sharing policies in developing countries	
10:45	Implementing a research data access policy in South Africa	Michael Kahn, CREST, University of Stellenbosch, South Africa
11:05	Access to Research Data and Scientific Information Generated with Public Funding in Chile	Patricia Muñoz, CONICYT, Chile
11:25	The Management of Health and	Leonard E. G. Mboera and

APPENDIX A   149

	Biomedical Data in Tanzania: The Need for a National Scientific Policy	Benjamin Mayala, National Institute for Medical Research, Tanzania
11:45	The data-sharing policy of the World Meteorological Organization: The case for international sharing of scientific data	Jack Hayes, World Meteorological Organization and NOAA, United States
12:05	Moderated Panel Discussion	*(all morning speakers)* Moderator: Farouk El-Baz, Boston University
12:30	*Lunch*	

Session Three: Compelling Benefits
Session Chair: <u>Barbara Andrews</u>, University of Chile

13:45	Examples of past successes	
	Developing the rice genome in China	Huanming Yang, Beijing Genomics Institute, China
	Data-sharing in astronomy	Željko Ivezić, University of Washington
	Sharing engineering data for failure analysis in airplane crashes: Creation of a Web-based knowledge system	Dan Cheney, Federal Aviation Administration
14:45	*Break*	
15:15	Examples where more data sharing could make a big difference	
	Integrated disaster research: Issues Around data	Jane Rovins, ICSU Integrated Research on Disaster Risk, China
	Understanding Brazilian biodiversity: Examples where more data-sharing could make the difference	Vanderlei Canhos, Reference Center on Environmental Information, Brazil
	Social statistics as one of the instruments of strategic management of sustainable development processes: Compelling examples	Victoria Bakhtina, International Finance Corporation
	Remote sensing and In Situ measurements in the Global Earth Observation System of Systems	Curtis Woodcock, Boston University
16:45	Moderated Panel Discussion	*(all 8 speakers)*
17:30	*Adjourn*	

Day Two:
Session Four: The Limits and Barriers to Data Sharing
Session Chair: <u>Roger Pfister</u>, Swiss Academy of Sciences and ICSU/CFRS

8:45	Introduction	
9:00	Consideration of barriers to data sharing	Elaine Collier, National Center for Research Resources,

Time	Topic	Speaker
		National Institutes of Health
9:20	Artificial barriers to data sharing – Technical aspects	Donald Riley, University of Maryland, United States
9:40	Scientific management and cultural aspects	David Carlson, University of Colorado[1], United States
10:00	Political and economical barriers to data-sharing: The African perspective	Tilahun Yilma, University of California, Davis, United States
10:20	Moderated Panel Discussion	(all 4 speakers) Moderator: Roger Pfister, Swiss Academies of Arts and Sciences and ICSU/CFRS, Switzerland
11:00	*Break*	

Session Five: How to Improve Data Access and Use
Session Chair: <u>John Rumble</u>, Information International Associates

Time	Topic	Speaker
11:20	Government science policy makers' and research funders' challenges to international data-sharing: The role of UNESCO	Gretchen Kalonji, UNESCO, France
11:40	International scientific organizations: Views and examples	Bengt Gustafsson, ICSU/CFRS, Sweden
12:00	Improving data access and use for Sustainable development in the south	Daniel Schaffer, TWAS, Italy
12:20	Lunch	
13:30	How to improve data access and use: An industry perspective	John Rumble, IIA, United States
13:50	Production and access to scientific Data in Africa: A framework for improving the contribution of research institutions	Hilary Inyang[2], African Continental University System Initiative, United States
14:10	The ICSU world data system	Yasuhiro Murayama, World Data System/NICT, Japan
14:30	Libraries and improving data access and use in developing regions	Stephen Griffin, National Science Foundation
14:50	Developing a policy framework to open up the rights to access and re-use	Haswira Nor Mohamad Hashim, Queensland University of Technology, Australia
15:10	Moderated Panel Discussion	
15:55	*Concluding observations*	Bengt Gustafsson, ICSU/CFRS, Sweden
16:00	*Adjourn*	

---

[1] Retired.
[2] Currently the Duke Energy Distinguished Professor of Environmental Engineering and Science, at the University of North Carolina, Charlotte. Dr. Inyang is past president of the African Continental University System Initiative.

Appendix B

Biographies of Symposium Chairs and Presenters

Charles M. Vest is President of the National Academy of Engineering and President Emeritus of the Massachusetts Institute of Technology. Dr. Vest earned a B.S. in mechanical engineering from West Virginia University in 1963, and M.S.E. and PhD degrees in mechanical engineering from the University of Michigan in 1964 and 1967 respectively. He joined the faculty of the University of Michigan as an assistant professor in 1968 where he taught in the areas of heat transfer, thermodynamics, and fluid mechanic, and conducted research in heat transfer and engineering applications of laser optics and holography. He and his graduate students developed techniques for making quantitative measurements of various properties and motions from holographic interferograms, especially the measurement of three-dimensional temperature and density fields using computer tomography. He became an associate professor in 1972 and a full professor in 1977.

In 1981 Dr. Vest turned much of his attention to academic administration at the University of Michigan, serving as associate dean of engineering from 1981-86, dean of engineering from 1986-1989, when he became provost and vice president for academic affairs. In 1990 he became president of the Massachusetts Institute of Technology (MIT) and served in that position until December 2004. He then became professor and president emeritus. As president of MIT, he was active in science, technology, and innovation policy; building partnerships among academia, government and industry; and championing the importance of open, global scientific communication, travel, and sharing of intellectual resources. During his tenure, MIT launched its OpenCourseWare (OCW) initiative; co-founded the Alliance for Global Sustainability; enhanced the racial, gender, and cultural diversity of its students and faculty; established major new institutes in neuroscience and genomic medicine; and redeveloped much of its campus.

He was a director of DuPont for 14 years and of IBM for 13 years; was vice chair of the U.S. Council on Competitiveness for eight years; and served on various federal committees and commissions, including the President's Committee of Advisors on Science and Technology (PCAST) during the Clinton and Bush administrations, the Commission on the Intelligence Capabilities of the United States Regarding Weapons of Mass Destruction, the Secretary of Education's Commission on the Future of Higher Education, the Secretary of State's Advisory Committee on Transformational Diplomacy and the Rice-Chertoff Secure Borders and Open Doors Advisory Committee. He serves on the boards of several non-profit organizations and foundations devoted to education, science, and technology. In July 2007 he was elected to serve as president of the U.S. National Academy of Engineering (NAE) for six years. He has authored a book on holographic interferometry, and two books on higher education. He has received honorary doctoral degrees from fourteen universities, and was awarded the 2006 National Medal of Technology by President Bush.

Barbara Andrews is Professor in the Department of Chemical Engineering and Biotechnology at the University of Chile in Santiago and a member of the Millennium Institute for Cell Dynamics and Biotechnology (ICDB). She has a BSc in Biochemistry and a PhD (Biochemical Engineering) from the University of London. Before joining the Department in Santiago 14 years ago, Barbara worked at Columbia University in New York and the University of Reading in the UK. Her main research interests include metabolic engineering, protein separations and purification, cold-adapted enzymes. She is a member of the ICSU Strategic Coordinating Committee on Data and Information (SCCID).

Victoria Bakhtina has over 10 years of experience devising portfolio and risk solutions and managing projects requiring collaboration across sectors, regions, and disciplines. She manages a corporate

Portfolio Screening initiative, addressing integrity of information for key portfolio areas, and leads an innovative Portfolio Management e-learning development. In addition, Dr. Bakhtina has worked in the areas of portfolio performance analysis (including applications to a corporate incentive program), credit research and rating performance, and loan recovery analysis. She has also built models for client financial reporting and compliance analysis. During the last five years, her scientific interests have been related to Risk Management, Sustainable Development, Knowledge Management, Innovation, and Leadership.

Dr. Bakhtina holds the equivalent of a Bachelors and Masters in Applied Mathematics and Composite Materials Mechanics from Lomonosov Moscow State University, a Ph.D. in Applied Mathematical Education from Moscow State Pedagogical University, a MBA in Business and Finance from Waynesburg University, and a certificate in Credit Risk Modeling from Stanford University. Ms. Bakhtina is an active member of CODATA Germany, and has a series of publications in the area of Sustainability and Developmental Risks.

Roberta Balstad is Co-chair of the Board on Research Data and Information of the National Research Council. She is also chair of the US National Committee on Scientific Data and Information (CODATA), a Trustee of the University Corporation for Atmospheric Research (UCAR), and a member of the Governing Board of the National Center for Earthquake Engineering Simulation. She is Editor-in-Chief of Weather, Climate, and Society, a journal of the American Meteorological Society, and a Special Research Scientist at Columbia University's Center for Research on Environmental Decisions. She has also served as Director of CIESIN (Center for International Earth Science Information Network), Columbia University, and as President/CEO of the organization before it moved to Columbia.

Dr. Balstad chaired the international committee that prepared a strategic plan on data and information for the International Council of Science (ICSU) and serves on that organization's Committee on Science and Policy Research (CSPR). Her publications focus on the role of the social sciences in understanding global environmental change, data policy and scientific research, information technology in science, and the social applications of remote sensing data. She was awarded the Ph.D. from the University of Minnesota in 1974. Dr. Balstad was previously the Director of the Division of Social and Economic Sciences at the National Science Foundation and the founding Executive Director of the Consortium of Social Science Associations (COSSA).

Vanderlei Canhos is the president director of the Reference Center on Environmental Information (CRIA) a non-governmental not-for-profit organization devoted to making biodiversity data freely and openly available via the Internet. Dr. Canhos has a B.S and a M.s. from the State University of Campinas (UNICAMP) and a Ph.D. degree from Oregon State University (OSU). His early scientific research was focused on microbial systematics at the State University of Campinas (UNICAMP). In the 1990's his work was focused on microbial collection management at the Tropical Culture Collection(CCT), and biodiversity databases at the Tropical Database(BDT).

During the last 10 years his work on the development of strategies and policies for the consolidation of biological collections and biodiversity databases in Brazil and Latin America. He contributed to the design and implementation of the Inter-American Biodiversity Information Network (IABIN), the Virtual Institute of Biodiversity (Biota-Fapesp Program), and in the implementation of the Brazilian Network of Biological Resource Centers. As a consultant to the Organization for Economic Cooperation and Development (OECD) he contributed to the development of the Best Practices Guidelines for the Operation and Management of Biological Resources Centers. He was the project leader of the speciesLink network (http://splink.cria.org.br/) and openModeler (http:// openmodeller.sourceforge.net/) a computational environment for ecological niche modeling. speciesLink currently integrates more than 4 million records from biological collections (botanical and zoological specimens and microbial strain data of samples collected in Brazil).

His current work includes the assessment of initiatives worldwide to improve data release and analysis. He is a member of the Board of Directors of Species2000 (Catalogue of Life), the Board of Directors of ETI Bioinformatics (University of Amsterdam) and of the Advisory Board of Global Research Data Infrastructures (GRDI 2020).

David Carlson has served as Director of the Atmospheric Technology Division of the US National Centre for Atmospheric Research (NCAR) Colorado for nine years between 1994-2003. Prior to that he was for three years the Director of the office coordinating the 'Tropical Ocean Global Atmosphere Coupled Ocean - Atmosphere Response Experiment' (TOGA COARE) leading oceanic and atmospheric scientists from twelve nations in a large ICSU/WMO climate research programme. He was appointed to the post of Director of the International Polar Year 2007-2008 International Programme Office (IPO) in June 2006. David is originally from the mid west of the United States and holds a PhD in Oceanography from the University of Maine.

Daniel I. Cheney is Manager of Safety Programs for the U.S. Federal Aviation Administration's Transport Airplane Directorate. Mr. Cheney has been involved in nearly all aspects of certification activities of Boeing commercial airplanes from the B707, through current activities of the B787 program. He has also been involved in supporting numerous accident and incident investigations, and their resolution. It has been through his involvement in accident investigation, and the realization that costly lessons from major accidents were being lost through the passage of time, that led Mr. Cheney to initiate FAA's development of the web-based "Lessons Learned from Transport Airplane Accidents" knowledge information system. This accident library, now containing 57 accident modules, is available to the public on the FAA's main web site at http://accidents-ll.faa.gov/.

Farouk El-Baz is Research Professor and Director of the Center for Remote Sensing at Boston University. He participated in NASA's Apollo program (1967-72) as secretary of the lunar landing site selection committee, and chairman of astronaut training in visual observations and photography. In 1973-82 he established and directed the Center for Earth and Planetary Studies at the National Air and Space Museum of the Smithsonian Institution, Washington DC and served as Principal Investigator of the Earth Observations and Photography Experiment on the Apollo-Soyuz mission of 1975.

He became Vice President for Science and Technology at Itek Optical Systems in 1982, before joining Boston University in 1986. He pioneered the applications of space photography to desert studies, with emphasis on groundwater concentration. He served in 1978-1981 as science adviser to Anwar Sadat, the late President of Egypt. His honors include NASA's Apollo Achievement Award, the Nevada Medal, and the Egyptian Order of Merit - First Class. He served for six years as chair of the U.S National Committee for the International Union of Geological Sciences of the National Academies, and is a member of the U.S. National Academy of Engineering.

Stephen M. Griffin is a Program Director in the Division of Information and Intelligent Systems at the National Science Foundation (NSF). For the period 1994-2004, Mr. Griffin managed the Special Projects Program that included the Interagency Digital Libraries Initiatives and the International Digital Libraries Collaborative Research and Applications Testbeds program. He has been active in working groups for Federal high performance computing and communications programs, and serves on numerous domestic and international advisory committees related to digital libraries and advanced computing and networking infrastructure. In 2004-2005 he was on special assignment to the Library of Congress, Office of Strategic Initiatives, to assist with the National Digital Information and Infrastructure Preservation Program. He is

currently again on assignment to the Library of Congress as a Visiting Scientist and advisor to the Associate Librarian for Strategic Initiatives and Senior Staff. His research interests include interdisciplinary scholarly communication, cultural heritage informatics and data-intensive scholarship.
Bengt Gustafsson is a Swedish astronomer and professor in theoretical astrophysics with stellar physics as is main are of research at Uppsala University. He is known for his work in uniting cosmic science with culture and theology, and questioning space science from a humanistic point of view. In 2002, Bengt was awarded the grand prize of the Royal Institute of Technology, and has also been awarded the grand prize of Längmanska kulturfonden. At one point during his career, he was a counselor working for the government of Sweden. He is a member of the Royal Swedish Academy of Sciences and the Norwegian Academy of Science and Letters.

Haswira Nor Mohamad Hashim is a PhD Candidate at the Queensland University of Technology (QUT) Faculty of Law, Brisbane, Australia. He holds a Bachelor and a Masters of Law degrees and is a legal academic at MARA University of Technology, Shah Alam, in his native country, Malaysia. Since 2010, Haswira has worked on his doctoral research project under the supervision of Professors Anne Fitzgerald and Brian Fitzgerald at QUT. His doctoral research is examining the policy framework supporting open access to and reuse of publicly funded research data and information in Malaysian public universities.

John L. "Jack" Hayes is the National Oceanic and Atmospheric Administration (NOAA) Assistant Administrator for Weather Services and National Weather Service (NWS) Director. In this role, Dr. Hayes is responsible for an integrated weather services program, supporting the delivery of a variety of weather, water, and climate services to government, industry, and the general public, including the preparation and delivery of weather warnings and predictions, and the exchange of data products and forecasts with international organizations.

Dr. Hayes returned to the NWS in 2007 after serving as the director of the World Weather Watch Department at the World Meteorological Organization (WMO), a specialized agency of the United Nations located in Geneva, Switzerland. In that position, he was responsible for global weather observing, weather data exchange telecommunications, and weather data processing and forecasting systems.

Before joining the WMO, Dr. Hayes served in several senior executive positions at NOAA. As the Deputy Assistant Administrator for NOAA Research, he was responsible for the management of research programs. As Deputy Assistant Administrator of the National Ocean Service (NOS), he was the chief operating officer dealing with a multitude of ocean and coastal challenges, including the NOS response to the Hurricane Katrina disaster in August 2005. As Director of the Office of Science and Technology for the NWS, Dr. Hayes had oversight of the infusion of new science and technology essential to weather service operations.

Dr. Hayes was also an executive in the private sector and the military. He was general manager of the Automated Weather Interactive Processing System (AWIPS) program at Litton-PRC from 1998 through 2000. AWIPS is the backbone computer and telecommunications system used by weather service forecasters at over 150 locations across the U.S. From 1970 through 1998, Dr. Hayes spent a career in the United States Air Force. He held a variety of positions, culminating his career as the Commander of the Air Force Weather Agency in the rank of Colonel.

Dr. Hayes received both his Ph.D. and Master of Science degrees in meteorology from the Naval Post Graduate School in Monterey, California. A Fellow in the American Meteorological Society, he also graduated from Bowling Green State University, with a bachelor's degree in mathematics.

Hilary Inyang is currently the Duke Energy Distinguished Professor of Environmental Engineering and Science, and Professor of Earth Science at the University of North Carolina, Charlotte, USA, while serving also as the Pro-term Chancellor and Board Chair of African Continental University Systems (ACUS) Initiative, Abuja, Nigeria. He is a former President of the African University of Science and Technology, Abuja, and Founding Director (2002–2009) of the Global Institute for Energy and Environmental Systems (GIEES) at the University of North Carolina-Charlotte. He is currently the President of the International Society for Environmental Geotechnology (ISEG) and the Global Alliance for Disaster Reduction (GADR). In 2008, he was selected as a Technical Judge of the US Nuclear Regulatory Commission. From 1997 to 2001, he was the Chair of the Environmental Engineering Committee of the United States Environmental Protection Agency's Science Advisory Board, and also served on the Effluent Guidelines Committee of the United States National Council for Environmental Policy and Technology. From 1995 to 2000, he was the DuPont Young Professor/University Distinguished Professor at the University of Massachusetts, where he helped establish the university System's Graduate School of Marine Science and Technology, while serving as the Founding Director of the Lowell-based Center for Environmental Engineering, Science and Technology. He has been an Honorary Professor/Concurrent Professor (China University of Mining and Technology, and Nanjing University) since 2004 and 1999, respectively.

He has authored/co-authored more than 220 research articles, book chapters, and federal design manuals and the textbook, Geoenvironmental Engineering: principles and applications, published by Marcel Dekker (ISBN: 0-8247-0045-7). He is the Editor-in-Chief of the Journal of Energy Engineering of the American Society of Civil Engineers (ASCE), and has been associate editor/editorial board member of 27 refereed international journals and contributing editor of three books, including the United Nations Encyclopedia of Life Support Systems (Environmental Monitoring Section). Professor Inyang has served on more than 100 technical and policy panels of governments and professional societies, and has given more than 120 invited speeches and presentations on a variety of technical and policy issues at many institutions and agencies in several countries, Professor Inyang holds a Ph.D. with a double major in Geotechnical Engineering and Materials, and a minor in Mineral Resources from Iowa State University, Ames, Iowa, USA. Prof. Inyang was the first African-American to be endowed as a distinguished professor in environmental engineering in any university in the United States, and the first African immigrant to chair a U.S Congress-chattered permanent science committee of any U.S federal agency He has received many professional prizes and awards.

Željko Ivezić is a professor in the Astronomy Department at the University of Washington. He holds a PhD in physics from the University of Kentucky, and B.Sc. degrees in physics and mechanical engineering from the University of Zagreb, Croatia. His research interests are in detection, analysis, and interpretation of electromagnetic radiation from astronomical objects, with emphasis on massive data sets and statistical analysis. He has co-authored over 250 refereed publications, with a cumulative citation count of over 40,000. He is one of the builders of the Sloan Digital Sky Survey (www.sdss.org), and serves as the Project Scientist for the Large Synoptic Survey Telescope (www.lsst.org).

Michael Kahn has contributed to science, education and innovation policy for over thirty years in strategy, policy, measurement, monitoring and evaluation of these fields. He has worked as advisor to the Ministers of Education, and Science and Technology, as Chief Director in the Gauteng Provincial Government, as education professor in Botswana and South Africa, as Executive Director in the Human Sciences Research Council, and as a consultant to many multilateral organizations. He presently serves as Research Fellow with the National Research Foundation, Honorary Research Fellow in the Centre for Research on Science and Technology of the University of Stellenbosch, and is Professor Extraordinaire of the Institute for Economic Research on Innovation of Tshwane University of Technology. Over 2010/11 he served as a member of the Ministerial Committee tasked to review South Africa's innovation system.

Gretchen Kalonji is the Assistant Director-General for Natural Sciences at the United Nations Educational, Scientific and Cultural Organization.

Throughout her career, she has developed strong international links in science, in particular in China, India and the Pacific Region. She is strongly committed to promoting science in Africa and has worked with several African universities. Her work in educational transformation has taken her to university positions in France, Japan and China. In 2006, she was appointed a distinguished honorary professor at Sichuan University, Chengdu, and a visiting professor at Beijing's Qinghua University.

Prior to 2005, she was the first women to hold an endowed chair – the Kyocera Professor of Materials Science and Engineering at the University of Washington (UW) Seattle, where she developed creative approaches to internationalization and to the transformation of science and engineering education. At UW, she led a campus-wide effort to integrate collaborative international research activities into the academic curriculum, across disciplines and from freshmen to doctoral level. This initiative, entitled UW Worldwide has been honored with multiple grants and awards, both in the United States and in partner regions.

Prior to 1990, when she joined the University of Washington, she served as Assistant and Associate Professor in the Department of Material Science and Engineering at the Massachusetts Institute of Technology (MIT).

Her work, both in materials science and in educational transformation, has been recognized by numerous awards and honors, including: the Presidential Young Investigator Award; the George E. Westinghouse Award from the American Society for Engineering Education; the Leadership Award from the International Network for Engineering Education and Research; and the National Science Foundation's Director's Award for Distinguished Teaching Scholars, the highest honor offered by the NSF. She has held visiting faculty appointments at numerous universities and institutes around the world, including the Max Planck Institute (Germany), the University of Paris (France), Tohoku University (Japan), and Sichuan University and Tsinghua University (China). She serves on numerous national and international advisory boards and committees, particularly for projects and organizations focusing on innovations in education, equity and access in higher education, and international science and engineering. She has been called upon to give more than 115 invited lectures in institutions around the world.

Benjamin K. Mayala attended the University of Dar es Salaam, Tanzania until 2002, obtaining a Bachelor of Science in Geomatics. He was employed by the National Institute for Medical Research (NIMR) in 2003 as a research scientist with expertise in health Geographic Information System (GIS). From 2005 to 2007, he pursued a Master of Science in GeoInformatics at the International Institute for GeoInformation Science and Earth Observation (ITC), Netherland.

Mr. Mayala has applied his GIS expertise by mapping the geographical locations of health facilities in Tanzania. He has worked on various health mapping and infectious diseases surveillance. He has worked with NGOs, government institutions and international organizations on various projects and research that utilise GIS. For the past 3 years, he was a mentor and lecturer on approaches in applying GIS in health research for Masters Studies in Field Epidemiology and Laboratory Technician at Muhimbili University of Health and Allied Science in Tanzania.

He is currently pursuing his Ph.D. at the University of Notre Dame in the Department of Biological Sciences. His project is focusing on the impact of climate on Change on Malaria in Tanzania, whereby he is going to use various algorithms to model the distribution of malaria in relation to climate variable. He is expecting to develop models that could be used as an early warning system for diseases prediction.

Leonard Mboera works with the National Institute for Medical Research in Tanzania Chief Research Scientist and Director of Information Technology and Communication. He was born in November 21, 1957. He holds a Bachelor of Veterinary Medicine of the Sokoine University of Agriculture (1985), MSc of Applied Entomology (Veterinary/Medical Entomology) of the University of London in UK and a Diploma of Imperial College, London, UK (1991). He obtained his PhD (Chemical Ecology) from the Wageningen University and Research Centre in the Netherlands in 1999.

He has worked as a Veterinary Surgeon at the Ministry of Agriculture, Zonal Research and Training Centre (1985-1992) before joining the National Institute for Medical Research in October 1992. In October 2002 he was appointed Director of Information Technology and Communication.

His scientific contributions include: development and improvement of mosquito sampling techniques for host seeking and oviposition site selections; mosquito human attractants and oviposition attractants; research on malaria epidemics and development of malaria early warning systems; ecohealth and linkages between agriculture and malaria; research on knowledge systems; and infectious disease surveillance systems.

Dr. Mboera is the Editor of Tanzania Journal of Health Research (2002- to-date); Associate Editor, East African Journal of Public Health (2004-to date); Member, Steering Committee, Global Outbreak Alert and Response Network (2003-2009); Coordinator, East African Integrated Disease Surveillance (2002-2004); and Secretary General, East African Public Health Association (2003- to-date). He is an author of one book, two book chapters, 72 peer reviewed Journal publications and 26 scientific technical reports.

Patricia Muñoz has a Bachelor of Information Sciences, and a Master's Degree in Digital Documentation at the Universidad Pompeu Fabra, Spain. She is the Director of the Scientific Information Programme at the National Commission of Scientific and Technological Research (CONICYT). She has specialized in Scientific and Technological Information Management, acting as the representative of Chile in various international experts committees, participating in projects of national and international scope focused on strengthening and guaranteeing accessibility and visibility of scientific and technological knowledge. She has been the representative of Chile in ASFA (Aquatics Science Fisheries Abstracts) BOARD), a database management by FAO. Since 2004, she has been a member of the Group of Experts in Marine Information Management (GEMIN), of UNESCO's International Oceans Commission.

She is currently the Director of the Scientific Information Programme at CONICYT, where she develops and leads projects related to the accessibility and visibility of projects and results of funding instruments managed by CONICYT's Programmes through the implementation of digital repositories with the purpose of guaranteeing accessibility and preservation of digital documents, and also implementing a portal with indicators of scientific productivity.

Yasuhiro Murayama is a director of Integrated Science Data System Research Laboratory of National Institute of Information and Communications Technology (NICT), Japan. He is in charge of WDS-IPO business in NICT and plays a role of liaison with ICSU and WDS-science committee. From 1999-2006 he had been a leader of Japan group of US-Japan joint program of Arctic middle-upper atmosphere observations in Alaska, and then he was appointed as a group leader of NICT's whole atmospheric remote sensing group in 2008-2011. He was awarded by Japan's Ministry of Education, Culture, Sports, Science and Technology in 2007, for Internet use of high-quality Auroral image observation and live image distribution. Dr. Murayama received his Ph.D. from Kyoto University studying atmospheric dynamics using remote-sensing and sounding rocket techniques.

Roger Pfister is Head of International Cooperation at the Swiss Academies of Arts and Sciences, the umbrella organization of the four Swiss academies. With an office at the Swiss Academy of Sciences (SCNAT), a member of this umbrella organization and the International Council for Science (ICSU), he is also the Executive Secretary for the ICSU Committee on Freedom and Responsibility in the conduct of Science (CFRS).

By formation, he is a political scientist and International Relations scholar. He did his graduate studies in contemporary history, with a focus on Africa, at the University of Berne, Switzerland, and at the Centre of West African Studies (CWAS), University of Birmingham, UK. He began his PhD at the Centre for International Studies (CIS) in Zurich, Switzerland, obtaining his degree from Rhodes University in Grahamstown, South Africa. In his doctoral thesis, he researched South Africa's apartheid foreign relations with sub-Saharan African states.

Following his studies, he was engaged in a network at the ETH Zurich that promoted the transfer of know-how to developing countries. Subsequently, in Fribourg, Switzerland, he directed the university's Research Promotion Service. In that capacity, he was also involved in setting up social sciences research programs in the Western Balkans and in the South Caucasus.

He has widely published on South Africa's post-Apartheid foreign policy, the diplomacy in exile of the African National Congress (ANC), Africa's international relations, and on the relevance of new information technologies for Africa.

Atta-ur-Rahman is the first scientist from the Muslim world to have won the prestigious UNESCO Science Prize (1999) in the 35 year old history of the Prize. He was elected as Fellow of Royal Society (London) in July 2006 thereby becoming the one of the 4 scientists from the Muslim world to have ever won this honor conferred by the prestigious 360 year old scientific Society. He has been conferred honorary doctorate degrees by many universities including the degree of Doctor of Science (Sc.D.) by the Cambridge University (UK) (1987) and an honorary degree of Doctor of Education by Coventry University UK in November 2007, an honorary D.Sc. degree by Bradford University in 2010, and an honorary Ph.D. by Asian Institute of Technology in 2010. He was elected Honorary Life Fellow of Kings College, Cambridge University, UK in 2007. He was conferred the TWAS Prize for Institution Building in Durban, South Africa in October 2009 in recognition of his contributions for bringing about revolutionary changes in the higher education sector in Pakistan.

He is President of Network of Academies of Sciences of Islamic Countries (NASIC) and the Vice-President (Central & South Asia) of the Academy of Sciences for the Developing World (TWAS) Council, and Foreign Fellow of Korean Academy of Sciences. He was the President of the Pakistan Academy of Sciences (2003-2006), and again elected the President of the Academy from 1st January 2011. He was the Federal Minister for Science and Technology (2000 to 2002), Federal Minister of Education (2002), and Chairman of the Higher Education Commission with the status of a Federal Minister from 2002-2008. Successive Governments of Pakistan have conferred four civil awards, including Tamgha-i-Imtiaz (1983), Sitara-i-Imtiaz (1991), Hilal-i-Imtiaz (1998), and the highest national civil award Nis-han-i-Imtiaz (2002), on him. The Austrian government also honored him with its high civil award (Grosse Goldene Ehrenzeischen am Bande" (2007) in recognition of his eminent contributions.

He is presently the Coordinator General of COMSTECH, an OIC Ministerial Committee comprising the 57 Ministers of Science & Technology from 57 OIC member countries. He is also the Patron of International Centre of Chemical and Biological Sciences (which comprises a number of institutes, including the Husein Ebrahim Jamal Research Institute of Chemistry and the Dr. Panjwani Center of Molecular Medicine and Drug Development) at Karachi University.

He obtained his Ph.D. in organic chemistry from Cambridge University (1968). He has over 843 publications in several fields of organic chemistry including 663 research publications, 18 patents, 103 books and 59 chapters in books published by major U.S. and European presses. Seventy four students have completed their Ph.D. degrees under his supervision.

Donald R. Riley is Professor of Information Systems, Robert H. Smith School of Business, and Affiliate Professor of Mechanical Engineering, at the University of Maryland, College Park. He also serves as IT Fellow at the Southeastern Universities Research Association (SURA) in Washington, D.C. He is founding chair of the Board of Directors of the Internet Educational Equal Access Foundation (IEEAF – http://www.ieeaf.org/), and serves on the board of National LambdaRail (NLR), Inc. Dr. Riley is co-founder and co-chair of the annual Chinese American Network Symposium and was recognized in 2000 by the Chinese Academy of Sciences as "Senior Technical Advisor to China Science and Technology Network."

Dr. Riley was one of the founding members of the national Internet2 initiative; founded the Mid-Atlantic Crossroads (MAX) regional networking consortium, one of the largest Internet2 regional gigapops, hosting the NGIX-DC (Next Generation Internet Exchange) for the federal agency NGI R&D networks. He served as inaugural chair of the EDUCAUSE Board of Trustees and was one of the founders of the EDUCAUSE National Learning Infrastructure Initiative.

From 1998 to 2003, Dr. Riley served as Vice President and CIO at the University of Maryland; from 1992 to 1998, he served in a similar capacity, Associate Vice President and CIO, at the University of Minnesota, where he also was a Professor of Mechanical Engineering from 1976 to 1998. Dr. Riley graduated from the Purdue University School of Mechanical Engineering with a B.S in 1969, an M.S. in 1970, and a Ph.D. in 1976.

Jane E. Rovins is an Executive Director for the Integrated Research on Disaster Risk Programme and International Programme Office. She has worked as a consultant promoting risk reduction, mitigation planning and providing emergency management training throughout Asia, USA, Latin America and Africa; was a Professor for American Military University in the Emergency and Disaster Management Department; and a Disaster Assistance Employee with the U.S. Federal Emergency Management Agency (FEMA) for numerous disasters. Her specialties are disaster risk reduction, programme management, planning, organizational development, and education.

She is a Certified Emergency Manager (CEM) and Florida Professional Emergency Manager (FPEM). Dr. Rovins is Scholarship Commission Chairman and Oceania/Asia CEM Commission Chair for the International Association of Emergency Managers (IAEM) and she is a Past-President of the Natural Hazard Mitigation Association and a member of the Board of Directors. She holds a Ph.D. in International Development from Tulane University Law School and is a graduate of Tulane University School of Public Health and Tropical Medicine with a Master's degree in International Health and Complex Emergencies.

John Rumble has over 30 years of experience managing scientific data organizations. He had nearly a quarter of a century of service with the National Institute of Standards and Technology (NIST) and its Standard Reference Data Program. From 2004 to 2001, He was senior executive for Information International Associates (Oak Ridge TN) and in 2011 founded R&R Data Services (Gaithersburg MD). Rumble's expertise in scientific data management and scientific informatics, as well as his broad scientific background, has expanded both company's breadth of experience and knowledge in managing scientific and technical information.

At NIST, Rumble led development of scientific and technical databases. Under his leadership, NIST produced and sold more than 70 PC databases of scientific information, generating millions of dollars of annual revenue. Dr. Rumble led NIST into the Internet and Web eras with the release of over 25 online data systems in virtually every area of science and engineering. When he left NIST, Rumble had oversight of direct customer service throughout the world as Chief of the NIST Measurement Services Division. The Division provides fee-based services to make instruments and measurements directly traceable to the fundamental units. It also provides critically evaluated databases in many areas of science and technology. During his time at NIST, he also worked closely with industry on database standards, including an international standard for industrial data exchange (ISO 10303). In 1998, he was elected President of the CODATA, the ICSU Committee on Data for Science and Technology), which is the leading international group for scientific and technical data.

In addition to his service to CODATA, Dr. Rumble is a Fellow of the ASTM International, a Fellow of ASM International, a foreign member of the Russian Federation Academy of Metrology, a Fellow of the International Union of Pure and Applied Chemistry, a Fellow of the American Association for the Advancement of Science, and recipient of the U.S. Department of Commerce Silver Medal. In 1993–94, Rumble was a Department of Commerce Fellow working in the Office of Science and Technology Policy in the Executive Office of the President. In 2006, Dr. Rumble was awarded the CODATA Prize for his accomplishments. He has written and lectured extensively on scientific data management. Dr. Rumble holds a Ph.D. in chemical physics from Indiana University.

Daniel Schaffer is a senior communications specialist with TWAS, the academy of sciences for the developing world, in Trieste, Italy, a UNESCO-affiliated organization dedicated to building scientific capacity in the developing world. He has written on science and technology issues both in the developed and developing world for more than 20 years. Earlier in his career, he served as director of communications at the University of Tennessee's Energy, Environment and Resources Center and an editor and speechwriter for the Tennessee Valley Authority (TVA) in the United States.

He has published books with Harvard University Press, Johns Hopkins University Press and World Scientific, and has produced educational programmes for WCBS-TV in New York City and award-winning documentaries aired on public television stations in the United States. He was the founding editor and editor-in-chief of Forum, a science policy journal published by the University of Tennessee, Tennessee Valley Authority (TVA) and Oak Ridge National Laboratory (ORNL). He holds a Ph.D. from Rutgers University, USA, where he studied issues related to urban and suburban growth.

Curtis E. Woodcock is a Professor in the Department of Geography and Environment at Boston University. His research specializations include remote sensing, particularly of land cover and land use change. He currently serves as the Team Leader of the USGS/NASA Landsat Science Team; Co-chairs the Land Cover Characteristics Implementation Team for GOFC-GOLD (Global Observations of Forests and Land Cover Dynamics); is a member of NASA's Land Cover and Land Use Change Science Team.

Huanming Yang co-founded BGI (formerly Beijing Genomics Institute) in 1999 and is currently its President and Professor. He and his collaborators have made a significant contribution to the Human Genome Project, the HapMap Project, and the 1000 Genomes Project, as well as to sequencing and analyzing genomes of the first Asian individual, rice, chicken, silkworm, giant panda, ant, cucumber, maize, soybean, and many microorganisms. Dr. Yang has received many awards and honors, including Research Leader of the Year by Scientific American in 2002 and Award in Biology by the Third World Academy of Sciences (TWAS) in 2006. He was elected as a Foreign Member of EMBO in 2006, an Academician of the Chinese Academy of Sciences in 2007, a Fellow of TWAS in 2008, and a Foreign Fellow of Indian National Science Academy in 2009.

Tilahun Yilma is the Director and Distinguished Professor of Virology at the International Laboratory of Molecular Biology (ILMB) at UC Davis. He received all his undergraduate, veterinary school, and graduate degrees at UC Davis. Dr. Yilma has made major contributions in developing safe and efficacious viral vaccines for both animals and humans. He is noted for developing what the journal Nature described as one of the two outstanding vaccinia virus recombinant vaccines (rinderpest) in the world.

Dr. Yilma has received many honors including the Ciba-Geigy award (highest international award in animal science), the UC Davis Faculty Research Award (highest research award), and UC Davis Distinguished Public Service Award (highest service award). Dr. Yilma is a member of the US National Academy of Sciences and a Fellow of the American Academy of Microbiology. Dr. Yilma has served on the Council of the NIH-NCRR, NSF Advisory Committee for the Office of International Science and Engineering, and currently is a Member of the Scientific Advisory Board for Center for Bio-Security Science, Los Alamos National Lab. Dr. Yilma has worked with a number of International Organizations including the WHO, FAO, OIE, AU, and IAEA. He is a member of the Board of Trustees, Science for Peace, Fondazione Umberto Veronesi, Milan, Italy.

Appendix C
CFRS Advisory Note[1]

International Symposium "The Case for International Sharing of Scientific Data: A Focus on Developing Countries"

Researchers in developing countries, in particular, lack the norms and traditions of more open data sharing for collaborative research and for the development of common research resources for the benefit of the entire research community. This international symposium was designed to help address these and related issues, and to improve the access to and use of publicly funded scientific data.

Synopsis

Scientific research and problem solving are increasingly dependent for successful outcomes on access to diverse sources of data generated by the public and academic research community. Global issues, such as disaster mitigation and response, international environmental management, epidemiology of infectious diseases, and various types of sustainable development concerns, require access to reliable data from many, if not all, countries. Digital networks now provide a near-universal infrastructure for sharing much of this factual information on a timely, comprehensive, and low-cost basis. There also are many compelling examples of data sharing in different areas that have yielded great benefits to the world community, although many more could be similarly facilitated.

Moreover, many OECD countries and some emerging economies already have implemented national policies and programs for public data management and access, while others are in the process of developing them. Nevertheless, a large number of developing countries do not have formal mechanisms in place.

At the same time, there are various specific barriers to the access and sharing of scientific data collected by governments or by researchers using public funding. Such obstacles include scientific and technical; institutional and management; economic and financial; legal and policy; and normative and socio-cultural aspects. Some of these barriers are possible to diminish or remove, whereas others seek to balance competing values that impose legitimate limitations on openness. Despite such challenges, however, there could be much greater value and benefits to research and society, particularly for economic and social development, from the broader use and sharing of existing factual data sources.

Researchers in developing countries, in particular, lack the norms and traditions of more open data sharing for collaborative research and for the development of common research resources for the benefit of the entire research community. Moreover, the governments in many developing countries treat publicly-generated or publicly-funded research data either as secret or commercial commodities. Even if the governments do not actively protect such data, they lack policies that

---

[1] See http://www.icsu.org/events/ICSUpercent20Events/international-symposium-the-case-for-international-sharing-of-scientific-date-a-focus-on-developing-countries.

provide guidance or identify responsibilities for the researchers they fund as to under what conditions the researchers should make their research data available for others to use. Finally, developing countries frequently do not have data centers or digital repositories in place that researchers can submit their data for use by others. In those cases where such repositories do exist, they tend to be managed as black archives.

This international symposium helped addressing these and related issues, and to improve the access to and use of publicly-funded scientific data.

Organizers and hosts

The symposium was organized by the Board on International Scientific Organizations (BISO), and the U.S. Committee on Data for Science and Technology (US CODATA) under the Board on Research Data and Information (BRDI), in consultation with the ICSU Committee on Freedom and Responsibility in the Conduct of Science (CFRS).